Bloomsburg University

Kendall Hunt
publishing company

Kendall Hunt
publishing company
www.kendallhunt.com
Send all inquiries to:
4050 Westmark Drive
Dubuque, IA 52004-1840

ISBN 978-1-4652-0084-6

Printed in the United States of America
10 9 8 7 6 5 4 3 2

CONTENTS

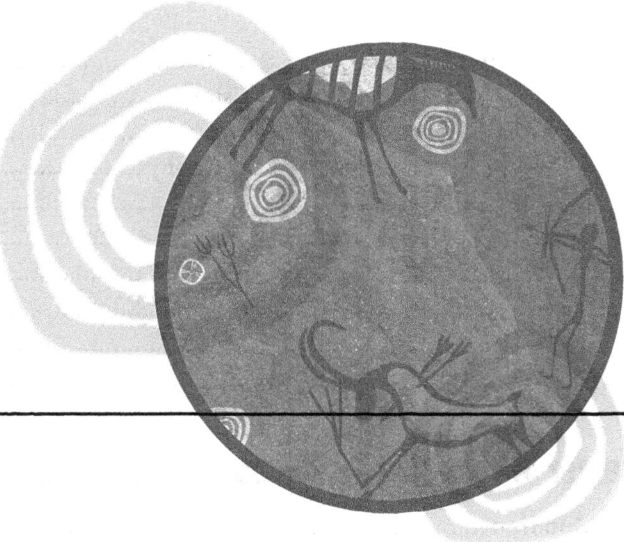

Introduction ix

Chapter 1: Physical (Biological) Anthropology 1

- Subfields in Physical Anthropology 1
  - Dead Populations 1
  - Living Populations 2
- What Is It to Be "Human"? 2
- Summary of Key Ideas 4
- Key Terms 4
- Further Readings 5
- Study Guide 5
- Home Work Exercises 7

Chapter 2: Evolutionary Theory 9

- Pre-Darwinian Thought 9
  - Middle Ages (Medieval—10th Century to 15th Century) 9
  - Age of Discovery (Renaissance—15th Century to 18th Century) 10
  - Age of Reason (Enlightenment—18th Century to 19th Century) 10
- Deep Time and the Beginnings of an Evolutionary Hypothesis 11
  - Geology: James Hutton and Charles Lyell 12
  - Darwin, Wallace, and the Theory of Natural Selection 14
- Summary of Key Ideas 16
- Key Terms 16
- Further Readings 17
- Study Guide 17
- Home Work Exercises 19

Chapter 3: Genetics and Heredity 23

- Molecular and Population Genetics 26
  - Molecular Level 26
  - Chromosomes 27
  - Mitosis and Meiosis 27

Protein Synthesis..28
Genetic (Blood) Polymorphisms..30
ABO Types and Disease History..32
Polygenic Variation and Pleiotrophy..32
Population Genetics..32
Summary of Key Ideas..33
Key Terms..34
Further Readings..35
Study Guide..35
Home Work Exercises..37

Chapter 4: Evolutionary Forces..43

Mutation..43
Natural Selection..43
Directional Selection..44
Stabilizing Selection..44
Balancing Selection..44
Genetic Drift..45
Gene Flow..46
Summary of Key Ideas..46
Key Terms..46
Further Readings..47
Study Guide..47
Home Work Exercises..49

Chapter 5: Primate Evolution..53

The Hypotheses for Primate Origins..53
Arboreal Hypothesis..53
Visual Predation Hypothesis..54
Angiosperm Radiation Hypothesis..54
Proprimates: First Primates?..54
Euprimates: The First True Primates..54
Anthropoids: The Higher Primates..55
Platyrrhine Origins: New World Monkeys..56
Branisella boliviana..56
Catarrhine Origins 1: Old World Monkeys..56
Aegyptopithecus zeuxis..56
Catarrhine Origins 2: Apes..57
Dryopithecines: The Mid-Miocene Apes..57
Summary of Key Ideas..59
Key Terms..60
Further Readings..60
Study Guide..60
Home Work Exercises..63

Chapter 6: Living Primates ... 67

Biogeography of Primates ... 67
General Primate Characteristics (Adaptations) ... 67
Taxonomic Classification ... 70
  Prosimians (Strepsirhini—Moist Nose) ... 70
  Anthropoidea (Haplorhini—Lack of Moist Nose) ... 70
Summary of Key Ideas ... 80
Key Terms ... 81
Further Readings ... 81
Study Guide ... 82
Home Work Exercises ... 83

Chapter 7: Dating Methods ... 89

Geological Time Period ... 89
  Paleozoic Era (545–245 Million Years Ago) ... 89
  Mesozoic Era (245–65 Million Years Ago) ... 89
  Cenozoic Era (65 Million Years Ago to Present) ... 90
Dating Methods ... 91
  Relative Dating ... 91
  Numerical Dating (Radiometric) ... 91
  Numerical Dating (Nonradiometric) ... 92
Reconstruction of Ancient Environments ... 92
  Stable Oxygen Isotope Studies ... 93
  Stable Carbon Isotope Studies and Animal Remains ... 93
Summary of Key Ideas ... 94
Key Terms ... 94
Further Readings ... 94
Study Guide ... 95
Home Work Exercises ... 97

Chapter 8: Australopiths and Early *Homo* ... 101

The Origin of Bipedalism ... 101
  Charles Darwin's Tool Use and Hunting Hypothesis for Bipedalism ... 101
  Energy Efficiency and Bipedalism Hypothesis ... 101
  Body Temperature and Bipedalism Hypothesis ... 102
  Male Provisioning Hypothesis for Bipedalism ... 102
Pre-Australopithecines ... 102
  Sahelanthropus tchadensis (Approximately 7 to 6 Million Years Ago) ... 102
  Orrorin tugenesis (Approximately 6 Million Years Ago) ... 103
  Ardipithecus ramidus (Approximately 5.8 to 4.4 Million Years Ago) ... 103
    Cranial and Dental Anatomy ... 103
    Postcranial Anatomy ... 103
Australopithecines ... 105
  East African Australopithecines ... 105
  South African Australopithecines ... 106
  Australopithecus sediba (Approximately 2.0 million years ago) ... 106

Robust Australopithecines (East and South Africa, Approximately 2.5 to 1.2 Million Years Ago)............ 108
Evidence of Tool Use in Australopiths? 109
Australopithecus garhi (2.5 Million Years Ago) 109
Australopithecus garhi at 2.5 million years ago and *Australopithecus sediba* at 2.0 million years ago: Implications for the Origin of Genus *Homo* 109
Early *Homo* 110
Homo habilis (East Africa, Approximately 2.0 to 1.8 Milion Years Ago) 110
Diets of Early Hominins 111
Dental Microwear Analysis 111
Stable Carbon Isotopes 113
Summary of Key Ideas 114
Key Terms 114
Further Readings 115
Study Guide 116
Home Work Exercises 117

Chapter 9: *Homo Erectus* 121

Pithecanthropus erectus 121
Morphology of *Homo erectus* 121
Geography 122
Homo erectus in Africa, Asia, and Europe (Approximately 1.8 Million to 300,000 Years Ago)............ 122
Tools and Subsistence 123
Speech and Language in *Homo erectus*? 123
Archaic *Homo sapiens* 124
Archaic *Homo sapiens* in Africa (Approximately 300,000 to 200,000 Years Ago) 124
Archaic *Homo sapiens* in Asia (Approximately 350,000 to 130,000 Years Ago) 124
Archaic *Homo sapiens* in Europe (Approximately 350,000 to 130,000 Years Ago) 124
Lithic Technology 125
Summary of Key Ideas 125
Key Terms 125
Further Readings 126
Study Guide 126
Home Work Exercises 127

Chapter 10: The Origins of Modern Humans 131

The Hypotheses for Modern Human Origins 131
Multiregional Continuity Hypothesis 131
Out of Africa Hypothesis 131
Assimilation Hypothesis 133
Neanderthals (130,000 to 24,000 Years Ago) 133
Geography 134
Neanderthal Biology 134
Neanderthal DNA 135
Culture and Behavior 135
Speech and Language in Neanderthals 135
Recent Discovery: The Denisovans (Approximately 50,000 to 30,000 years ago) 135

Other Early Moderns 136
Asia and Europe (Approximately 90,000 to 15,000 Years Ago) 136
Lithic Technology 137
The Cultural Revolution: The Shift to Agriculture 137
Peopling of the "New World" 137
Asian Ancestry for Native Americans: The Dental and Genetic Evidence 138
Paleoindians: Culture and Behavior 139
Summary of Key Ideas 139
Key Terms 140
Further Readings 140
Study Guide 141
Home Work Exercises 143

Chapter 11: Modern Humans: Climate, High Altitude, and Cultural Adaptation 147

Human Adaptation 147
Types of Adaptation 147
Genetic Adaptation 147
Developmental Adaptation 147
Physiological Adaptation 1 (Acclimation) 147
Physiological Adaptation 2 (Acclimatization) 148
Cultural Adaptation 148
Climate Adaptation 148
Cold Stress 148
Heat Stress 149
Skin Color: Pigmentation and Depigmentation 149
High Altitude Adaptation 150
Biocultural Adaptation 150
Nutrition and Disease 151
Summary of Key Ideas 152
Key Terms 153
Further Readings 153
Study Guide 153
Home Work Exercises 155

Glossary 159
References 169
Index 173

INTRODUCTION

Physical (biological) anthropology is the study of biological variation (which includes genetics) and evolution in living and dead human populations. In living populations, basic changes, such as body size, height, and so on, are seen in the next few generations. In dead populations, that is, prehistoric and historic peoples, changes are documented by studying the skeletal remains. Whether we are studying living or dead human populations, the framework of study is evolutionary. This means that changes in morphological (body) characteristics occur over time as populations are shaped, biologically and behaviorally, by their environment in a type of biocultural interaction. New species are formed, particularly in animal species, when the change occurs over long periods of time. While this seems clear to this author, the average student finds long-term evolutionary change and the forces causing the change too abstract to grasp. In fact, some questions posed by my students over the years are as follows:

1. If God did not create humans, where did the races come from?
2. If chimpanzees and humans share a 97–98 percent genetic similarity, shouldn't we look similar?
3. If humans originated in Africa, why are there so many different races?
4. Why are there so many hypotheses for the human family tree? Do anthropologists really know anything about human origins?

These questions are pertinent and difficult to answer. But, in terms of race—which is a difficult subject on its own—humanity did originate in sub-Saharan Africa based on the earliest prehominin and hominin fossils (evidence for bipedalism) and spread to different parts of the Old World (Middle East, Europe, Asia, and Indonesia) by 1.0 million years, adapting in their respective environments but remaining connected through gene flow (mating behavior), that is, alliances (exogamy—marriage) and warfare (rape and pillage). In essence, we (Africans, Europeans, Middle Easterners, Asians, Indonesians, etc.) have inherited our ancestors' traits: hair form, nose shape, and skin color.

Regulatory genes, important genes that turn other genes on and off, are responsible for the differences in structural morphology between humans and chimpanzees, so that the 97–98 percent genetic similarity is not important. With advances in technology and more primate and hominin fossil discoveries, we will be able to further illuminate the human–chimpanzee evolutionary relationship.

Compared to applied courses like forensic anthropology or human genetics, for example, human origins is a very difficult subject for the average student to master because of the complex evolutionary hypotheses, abstract evolutionary forces, and numerous fossil species. In this text, I will try to reduce this complexity by giving students a basic working knowledge of human origins without sacrificing key information. As such, students will learn the history of evolutionary theory; the evolutionary forces driving evolutionary change (selection, mutation, gene flow, and genetic drift); basic genetics and heredity; the living primates; and the biological and cultural variation of early, middle, and late Pleistocene (2.0 million to 10,000 years ago) hominins (*Australopithecines, Homo habilis, Homo erectus,* Neanderthals, and early modern humans). In the end, students will gain knowledge of (1) the history and development of evolutionary theory and the evolutionary forces that cause change; (2) the biology and behavior of nonhuman primates and how this behavior gives us clues to the earliest hominin behavior; and (3) how the earliest hominin ancestors came to be "human" and how culture and environment interacted to produce modern human anatomy.

CHAPTER 1

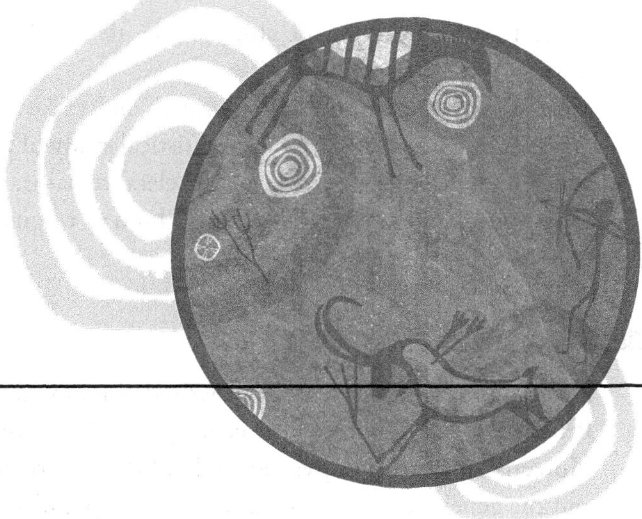

Physical (Biological) Anthropology

Physical (biological) anthropology is the study of biological variation (which includes genetics) and **evolution** in living and dead human populations all over the world. While the focus is biological, we take into account the cultural and environmental effects that impact biology. In essence, we approach **physical anthropology** from an evolutionary and **biocultural** perspective. Before 10,000 years ago, there is evidence in the fossil record for increased cranial capacity, reduction in the face and teeth, and complexity in **culture** and technology. From this we can infer complexity in behavior, that is, scavenging, social organization, cooperative hunting, myth and ritual, and language. There was less impact on the environment in terms of land use and animal extinction. After 10,000 years ago, the agricultural revolution (global warming forced humans to shift from hunting-and-gathering to agriculture) led to population increase, towns, cities, states, nutritional changes, and diseases because of high population densities in towns and cities. Infectious diseases, for example, were at a very low frequency before 10,000 years ago because hunter-and-gatherer populations were too small to sustain epidemics. After 10,000 years in the advent of agriculture, which includes animal and plant domestication, the frequency of infectious diseases increased. In short, the impact on the environment has steadily increased so that today we have global warming, soil erosion, urban development, and animal extinction. In addition, the evolutionary forces shaping human biology have drastically changed.

In the first paragraph, a few subfields in physical anthropology were indirectly discussed, such as human paleontology, paleopathology, and human biology. Now, let us directly define these subfields and include a few others.

Subfields in Physical Anthropology

DEAD POPULATIONS

Human paleontology: the study of human origins/evolution through the analysis of fossilized primate skeletal remains *showing evidence for bipedalism.*

Primate paleontology: the study of the evolution of the nonhuman **primates** through the analysis of fossilized skeletal remains.

Human paleopathology: the study of disease and trauma in prehistoric populations through the analysis of skeletal and dental remains.

Forensic anthropology: the study of trauma on modern human skeletal remains and the estimation of race, sex, age, and stature (height) in suspicious deaths.

Dental anthropology: the study of the dental development, variation, and evolution in prehistoric and historic skeletal populations.

Human osteology: the study of growth and variation through the analysis of skeletal remains.

Human paleodemography: the statistical study of prehistoric population size, distribution, density, births, and deaths through the analysis of skeletal remains.

Human paleogenetics: the study of human origins/evolution and population relationships by obtaining bone samples for genetic analysis.

LIVING POPULATIONS

Human biology: the study of growth, development, variation, nutrition, illness, and disease in populations.

Primatology: the study of the ecology, social structure, biology, and behavior of nonhuman primates.

Molecular anthropology: the study of the genetic distance between humans and the nonhuman primates and the genetic relationship among human populations.

The subfields of physical anthropology are varied, interesting, and complex. There is a traditional misconception—among students and some professionals—that the social sciences in general or anthropology in particular are nonscientific. This is wrong. In these subfields, data collection and analysis are based on the **scientific method**. In other words, a tentative statement about a phenomenon is made. This is known as a **hypothesis**. Subsequently, data are collected to test this hypothesis. The data are analyzed through observation and measurement. Finally, the hypothesis is accepted or rejected based on the results of the analysis.

Furthermore, students find the numerous hypotheses, uncertainties, and inconsistencies among professionals in human origins irritating, which leads most students to conclude, "They (paleoanthropologists) don't know what they are talking about." This is a false conclusion. Students must understand that there are no simple or perfect answers in science. "God created man in his own image"; this is a perfect answer. There is no irritation here; it does not require you to think. In contrast, science is tentative. This is why we have "**theories**" and not "fact." For instance, the theory of gravity and the theory of relativity are supported by tremendous evidence. Yet they are still theories. Why? Because a future scientist might come along in an age of more advanced technology and propose a more accurate explanation. For example, Newton had great anxiety that his new equations for gravity, which described the force of attraction between pairs of objects, might not maintain a stable system of orbits for several planets, causing them to crash into the sun or get ejected from the solar system (Tyson 1999, 2005). Worried about the long-term fate of Earth and the other planets, Newton invoked a Supreme Being as the force restoring order in this chaotic system. The French mathematician Pierre Simon de Laplace, a century after Newton, invented a mathematical approach to gravity, which extended the applicability of Newton's equations to complex systems of planets, and theorized that there was neither instability nor a "Supreme Being" in our solar system (Tyson 1999, 82; 2005, 28–29). In essence, science is self-correcting; there will always be hypotheses, theories, inconsistencies, and uncertainties. As such, human origins will always cause you some irritation. But be patient; the benefits of learning how humans came to be is worth this irritation.

What Is It to Be "Human"?

Are we unique? Students reading this textbook would answer "yes" because you see science and technology all around you and the absence of the same in the animal world. Even more obvious is the fact that humans communicate using language and speech, while animals use inchoate vocalizations. But long-term nonhuman primate studies (de Waal 1995, 1999; Fox and bin'Muhammad 2002; Goodall 1986; McGrew 1992, 1998; Mercader et al. 2002; Pruetz and Bertolani 2007; Savage-Rumbaugh and Lewin 1994; van Schaik and Knott 2001; Van Schaik et al. 2003) have found evidence of cultural behavior, that is, tool use and the transmission of that behavior to the offspring, particularly in the great apes (gorillas, orangutans, and chimpanzees). Also, some great apes are capable of learning American Sign Language (ASL) and language based on abstract symbols (Biro and Matsuzawa 2001; Gardner and Gardner 1969; Linden 1981; Patterson 1978; Patterson and Linden 1981; Savage-Rumbaugh and Lewin 1994; Savage-Rumbaugh et al. 1998; Snowden 1990). Since these behaviors threaten human uniqueness, unconscious bias has caused some researchers to dismiss these behaviors as a reflection of unconscious cues given by these animals' trainers.

In terms of tool use, some great apes make and use tools, for example, rocks to break open nuts, modified sticks to "fish" for termites, or leaves used as a "sponge" to soak up water to drink. But they have to make the respective tool every time they need to use it. In other words, they do not save their tools. Humans, in contrast, manufacture and *depend* on complex tools and technology, such as automobiles, computers, and airplanes, to name a few. We save this technology and use it to make more and better tools. In this sense and the fact that apes will never be able to speak or understand grammar and construct sentences, humans are unique.

Continuing, our big brain, smaller jaws and teeth (particularly our canines), and obligate bipedalism also make us unique. But how did this all come to be? Was it God using evolution to shape organisms in their respective environments? Or was it nondirected evolutionary forces working to shape organisms in their respective environments? Let us discuss the first question, in which the definition of science becomes critical.

In a recent collaborative effort, Baylor University researchers and the Gallup Organization conducted a national scientific survey on religious attitudes (Elmore 2007). The data revealed that 95 percent of the American population believed that a "God" does exist and that Americans are more religious than previous surveys suggested. This survey was conducted from October to December 2005 with 1,721 respondents, with a margin of error of +/−4 percent. Furthermore, of that 95 percent, 24.4 percent state that they believe God is distant—not active in the world, but a cosmic force that set the laws of nature in motion.

The idea of a Supreme Being or a God directing and shaping organisms is prominent in every human culture. This belief is functional in human societies because it helps us to cope in times of crisis. But how do we test the **God hypothesis**? Or can it be tested scientifically? Creationists—those individuals who believe in the sudden creation of all life by a creator—believe that it can. This is why they call their proposal "**creation science**." For any hypothesis to be scientific, it must be testable and subject to rejection. Creation science or the God hypothesis fails on both accounts. In fact, there is no "science" in creationism. It is simply "a grab bag of ideas spruced up with scientific jargon" (Relethford 2005, 24). A good example of this is **intelligent design creationism**. Supporters of this view claim that many of the characteristics of biological organisms, such as the structure of DNA or the anatomy of the eye, are too complex to have occurred through natural processes (or chance). Consequently, there must be a creator. Most students will not understand how a complex trait like the eye could "just" evolve. If it did, some would use the astrophysicist Sir Fred Hoyle's purported sarcastic comment that it would be as likely as "a hurricane having the luck to put together a Boeing 747 when whirling through a junk yard" (Dawkins 2006, 113). I must confess that I did not find the former quote in any of Hoyle's writings on evolution. But I did find his statement on the impossibility of monkeys producing works of Shakespeare by randomly hitting the keys of a typewriter (Hoyle and Wickramasinghe 1981, 16, 148):

> . . . a random copying error can lead to an appreciable improvement. . . . So the improvement tends to spread through the positive natural selection that is the basis of Darwinism. However, finding improvements to such a highly sophisticated molecule as hemoglobin is like monkeys on a typewriter finding improvements to Shakespeare. It can happen only very rarely, and the question is whether it can happen often enough to give Darwinism any hope of being able to explain the evolution of plants and animals. . . . No matter how large the environment one considers, life cannot have had a random beginning. Troops of monkeys thundering away at random on typewriters could not produce the works of Shakespeare, for the practical reason that the whole observable universe is not large enough to contain the necessary monkey hordes, the necessary typewriters, and certainly the waste paper baskets required for the deposition of wrong attempts. The same is true for living material.

A "**Grand Designer**" creating the eye for a specific purpose makes more sense. However, nondirectional, random, and adaptive cumulative changes over long periods of time lead to amazing results. While the God hypothesis cannot be tested or rejected, the evidence for evolution, however, has been tested and not rejected. Of course, some predictions of evolutionary theory have been rejected, but this is the nature of science. There is always error in science. Creationists attack the uncertainties in evolutionary theory and present in its place perfect answers, such as sudden creation of the universe, energy, and life from nothing; the occurrence of a worldwide flood; separate ancestry for man and apes; and a recent Earth. In the following chapters, students will learn about evolutionary theory and the evolutionary forces behind long-term cumulative changes. Also, you will learn very quickly that there are no easy answers in the origins of life in general and humans in particular.

Summary of Key Ideas

1. Physical anthropology is the study of biological variation and evolution in living and dead human populations all over the world, past and present.
2. Before 10,000 years ago, hunter-and-gatherer societies made very little impact on the environment, and these populations were too small to sustain epidemics.
3. After 10,000 years ago, the agricultural revolution (plant and animal domestication) led to population increase, towns, cities, states, nutritional changes, and infectious disease epidemics.
4. The subfields of physical anthropology study both living and dead human populations. Some of these major subfields are human paleontology, primate paleontology, human paleopathology, forensic anthropology, dental anthropology, human osteology, human paleodemography, human paleogenetics, human biology, primatology, and molecular anthropology.
5. The scientific method begins with a hypothesis—a tentative statement about the phenomenon in question. This hypothesis is tested by the collection, experimentation, and measurement of the data. Finally, the hypothesis is accepted or rejected. Any hypothesis that has been tested independently by several scientists and not been rejected becomes a theory.
6. Creation "science" is not science simply because the existence of God is not amenable to the scientific method. In other words, it cannot be tested or rejected, and therefore it is not science.
7. To be human is to depend on complex culture. Humans make and use complex tools; we save our tools and use our tools to make other tools. In addition, humans have obligate bipedalism and complex language and speech. Some animals are capable of making and using tools and learning American Sign Language. However, these behaviors are very simple.

Key Terms

Biocultural
Creation science
Culture
Dental anthropology
Evolution
Forensic anthropology
God hypothesis
Grand Designer
Human
Human biology
Human osteology
Human paleontology
Human paleopathology
Hypothesis
Intelligent design
Molecular anthropology
Physical anthropology
Primates
Scientific method
Theory

Further Readings

Futuyama, D. J. 1995. *Science on trial: The case for evolution,* rev. ed. Sunderland, MA: Sinauer Associates.

Moore, J. A. 1999. *Science as a way of knowing: The foundations of modern biology.* Cambridge, MA: Harvard University Press.

Pennock, R. T. 1999. *Tower of Babel: The evidence against the new creationism.* Cambridge, MA: MIT Press.

Spencer, F. 1997. *History of physical anthropology: An encyclopedia.* New York: Garland.

INTERNET RESOURCES

American Association of Physical Anthropologists
http://physanth.org/association
This Web site provides information on physical anthropology and the subfields, annual meetings, job announcements, graduate programs, and job openings.

National Center for Science Education
http://www.natcenscied.org
This Web site provides information on the continuing controversy over teaching creation science in the public schools.

Chapter 1: Physical (Biological) Anthropology

Study guide: By knowing the material below you will be well prepared for the exam. Don't forget to read the chapter and study your notes (this includes the chapter worksheets).

1. Be able to list and define the subfields of physical anthropology
2. Be able to define *Intelligent Design Creationism*
3. Be able to differentiate between a hypothesis and a theory
4. Be able to define human paleontology
5. Be able to explain the scientific method
6. Be able to explain the biocultural approach. Give an example
7. Know the five steps to humanness and be able explain each step

CHAPTER 2

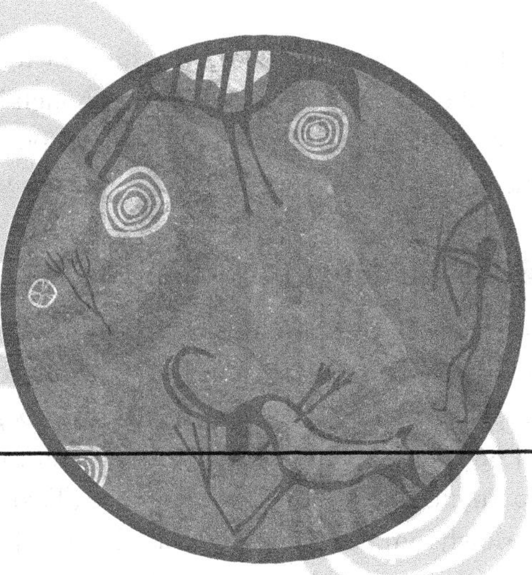

Evolutionary Theory

Pre-Darwinian Thought

Status, ranking, and the ideas of the "supernatural" and "perfection" have been a part of the human culture/psyche since the days of bands, tribes, and chiefdoms. And these concepts continue, slightly modified, in modern societies. For modern Western Judeo-Christian societies, one can trace the origins of these ideas in the Bible. In Genesis 1:26–28 of the King James Version God said:

> **26** Let us make man in our own image, after our likeness: and let them have *dominion over the fish of the sea, and over the fowl of the air, and over the cattle, and over all the earth, and over every creeping thing that creepeth upon the earth* [italics, my emphasis].
>
> **27** So God created man *in his own image*, in the image of God created he him; male and female created them [italics, my emphasis].
>
> **28** And God blessed them, and God said unto them, be fruitful and multiply, and replenish the earth, *and subdue it* [italics, my emphasis]: and have dominion over the fish of the sea and the fowl of the air, and over every living thing that moveth upon the earth. (Baker Publishing Group 1995, 1–2)

This biblical assertion was a persistent theme in Western societies, which was passed down unmodified for generations emphasizing dominance, and therefore distance, over the animal world.

MIDDLE AGES (MEDIEVAL—10TH CENTURY TO 15TH CENTURY)

In this cultural period, human beings believed in an unchanging world. All biological life on Earth had been present since the Creation, so nothing could be added (evolution) and nothing could be taken away (extinction). All aspects of life were fixed, which is referred to by some as the "**Fixity of Species**" (Jurmain and Nelson 1994). The general idea was that *God* was perfect, and *he* created a world in which everything was perfect. In essence, the concept of change was inconceivable because this would mean that *God* was imperfect. Moreover, all life was linked in a "great chain," which is commonly known as the **Great Chain of Being** or Aristotle's *scalae naturale* (Brace and Montagu 1977; Jurmain and Nelson 1994). This model went as far as proposing a hierarchy of life, which was roughly similar to the one listed below:

God
Angels
Humans
Animals
Inanimate objects

This hierarchy was fixed with everything, including human social organization, set nicely in its own place. The earth was seen as the center of the universe even though Nicolaus Copernicus (1473–1543) proposed a **heliocentric** solar system (Sakolsky 2005). At this time, hypotheses proposing environmental or biological change could not be formulated because there was no context for these ideas in a worldview of no change or **stasis**.

AGE OF DISCOVERY (RENAISSANCE—15TH CENTURY TO 18TH CENTURY)

In this period of exploration, ship travel was the primary mode of transportation. Of course, this was not the first time ships were used for travel. What changed was that ships in this period were built for travel over vast oceans. As Europeans explored new continents, they began to realize that the world was not "fixed." In essence, they discovered new varieties of plants, animals, and, to their surprise, people. Also, Galileo Galilei (1564–1642) confirmed Copernicus's heliocentric solar system (Tyson 2005). Despite the fact that the worldview of the "fixity of species" was slowly unraveling, the Great Chain of Being remained as a resistant and powerful belief (Brace 1996; Gould 1994; Jurmain and Nelson 1994).

AGE OF REASON (ENLIGHTENMENT—18TH CENTURY TO 19TH CENTURY)

The scientific revolution did not happen all at once. It probably had roots going all the way back to the Renaissance. Nonetheless, a generation of "giants," for example, Francis Bacon, Isaac Newton, Galileo Galilei, and Johannes Kepler, led this revolution. They had an amazing way of balancing science and religion by making religion the primary impetus for the study of natural philosophy, which today is known as science. Other writers (Jurmain and Nelson 1994) have stated that these scientists studied the natural world to show the glory of God. But it was Carolus Linnaeus (1707–1778), using Aristotelian logic, who devised a system that he believed created "order" out of the chaos of organic life (Brace 2005, 25).

Linnaeus was an Enlightenment figure who believed that the act of naming plants and animals was equated with showing the glory of God. This can be clearly seen in the words he used in the introduction to all the later editions of his *Systema Naturae*: "I saw the infinite, all-knowing and all-powerful God from behind as He went away and I grew dizzy. I followed his footsteps over nature's fields and saw everywhere an eternal wisdom and power, an inscrutable perfection" (Lindroth 1973, 380). In this respect, Linnaeus reflected the worldview of the medieval scholastics. He was also a complete Platonic essentialist and a classic creationist who believed that all biological species were fixed and mere imperfect reflections of their "true" underlying essences, and these true essences could only be found in the mind of God (Brace 2005, 27). According to this worldview, categories of species were not only treated in terms of similarity or difference, but were ranked based on the proximity or distance from the perfection of the divine. This arrangement was united with biblical assertions that human beings, as mentioned in the beginning of this chapter, were superior to all other forms of life and were given dominion over them by divine decree. This assertion, once more, is further emphasized in Psalms 8:4–9:

4 What is man that thou art mindful of him? And the son of man, that thou visitest him?
5 For thou hast made him *a little lower than the angels* [italics, my emphasis], and hast crowned him with glory and honour.
6 Thou madest him to have dominion over the works of thy hands; thou hast put all things under his feet:
7 All sheep and oxen, yea, and the beasts of the field;
8 The fowl of the air and the fish of the sea, and whatsoever passeth through the paths of the seas.
9 O LORD our Lord, how excellent is thy name in all the earth. (Baker Publishing Group 1995, 584)

The assumption that the world was ranked and fixed pervaded the medieval Christian worldview, and this worldview was inherited by Enlightenment thinkers. Linnaeus and his contemporaries simply took that general view and provided a more specific picture of all aspects of life arranged in a series of steps running from God at the top down through various entities of the living world to the inorganic—"base" metals—at the bottom (Brace 2005, 28). As Arthur O. Lovejoy (1961, 184) notes, "Next to the word 'Nature,' 'the Great Chain of Being' was the sacred phrase of the eighteenth century, playing a part somewhat analogous to that of the blessed word 'evolution' in the late nineteenth." In this Great Chain of Being anything closest to the divine had high status and embodied Plato's "Idea of Good." Plato is affirming, in what he says about the Idea of Good or *ens perfectissimum*, that a rational divine power exists and it is omnipotent (Lovejoy 1961, 40–42). For those who viewed the world as God-created, fixed, and perfect, ideas of variation, random change, or extinction were inconceivable. The loss of one of the steps in that Great Chain would mean that the world was not created perfectly or God was not all-powerful (Lovejoy 1961, 256–257). Of course, neither possibility was consistent with Enlightenment thinking: God was perfect, God did not make mistakes, and nothing occurred by chance.

Deep Time and the Beginnings of an Evolutionary Hypothesis

The name Charles Darwin (1809–1882) (Figure 2.1) is celebrated by evolutionists and cursed by creationists. Some have even referred to his book ***On the Origin of Species,*** published in 1859, as "the evolutionist's bible." This would be an incorrect reference. To give students a more accurate understanding of Darwinian evolution, Darwin's early life must be explored to learn how he proposed his mechanism for evolution.

Darwin was born into the Age of Reason. At this time, a few European scholars had heretical notions of biological variation and change and the ancient formation of the earth—different from that written in the Book of Genesis in the Bible. Darwin would eventually use their writings to generate his theory of evolution.

Figure 2.1. Charles Darwin in later years.

As a student at Cambridge University in England, Darwin turned away from two of the most prestigious career paths for young men in the early 19th century: the priesthood and medicine. He was, however, passionately interested in natural history and in collecting insects, birds, flowers, and rocks. In effect, he trained to become a naturalist. For an upper-middle-class, 19th-century young Englishman, this nonpractical career was not a major problem because his family had enough money so that he could live in a certain level of comfort without working. Nonetheless, through family connections, Darwin got the opportunity to sail as an unpaid naturalist aboard the survey ship HMS *Beagle* on a five-year journey around the world (see Figure 2.2).

This voyage was critical for Darwin because it enabled him to develop his theory of natural selection. Spending three-and-a-half years surveying and collecting along the coasts of South America, five weeks in the Galapagos Islands of Ecuador, and a year returning home via Tahiti, New Zealand, Australia, and South Africa enabled Darwin to see a tremendous amount of variation in most living species (Campbell et al. 2006; Relethford 2005). This variation was, in general, no different from what 15th- and 16th-century European explorers saw and wrote home about. Unlike these early explorers, Darwin would present this variation to the public in a scientific context.

Darwin observed that organisms seemed well adapted to their environments. For instance, he noted that organisms in cold climates have fur and birds in areas where seeds are hard have large beaks. One of the most important parts of the voyage for his understanding of **adaptation** was the visit to the **Galapagos Islands**. Darwin (1962 [1839]) wrote,

> It was most striking to be surrounded by new birds, new reptiles, new shells, new Insects, new plants, and yet by innumerable trifling details of structure, and even by the tones of voice and plumage of the birds, to have the temperate plains of Patagonia, or the hot dry deserts of northern Chile, vividly brought before my eyes. (393)

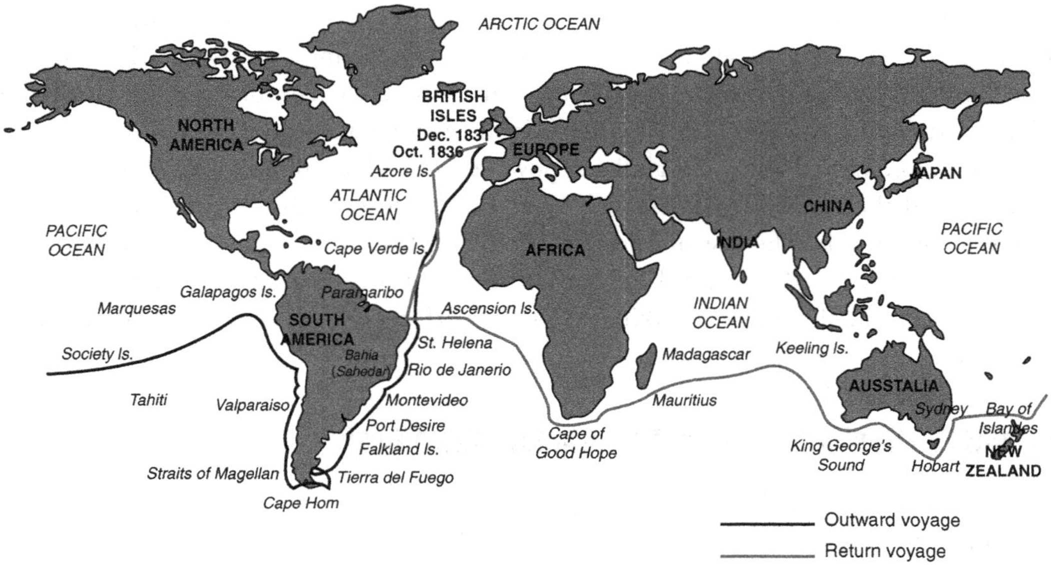

Figure 2.2. Darwin's five-year voyage and the different regions he visited aboard the HMS *Beagle* (from Relethford 2005).

The Galapagos animals were certainly different compared to the mainland species, but there were some similarities. In fact, Darwin believed that the finches on the islands originated from a common ancestor on the mainland (see Figure 2.3). Due to unknown circumstances, some individuals got separated on the respective islands and over time became separate species as they adapted to their respective niches. For example, on island to island, the finches' beaks were strong and thick for cracking big nuts and seeds; smaller for catching insects; and elongated for feeding on flowers and fruit (Campbell et al. 2006). In the initial arrival on the islands many generations ago, the finches had found different foods on the islands. Over time, they adapted to better survive in their respective environments.

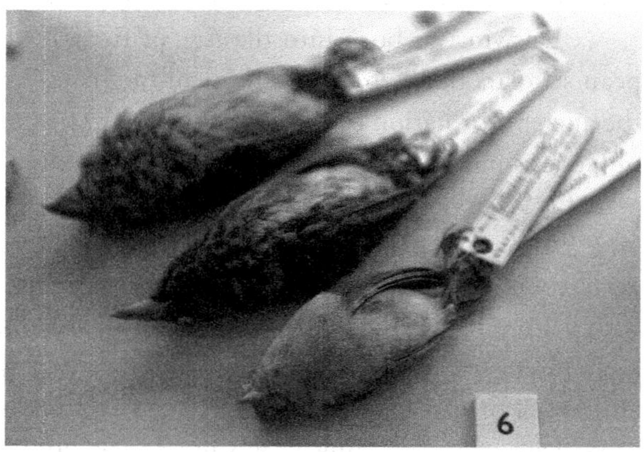

Figure 2.3. Three of Darwin's finches from the Galapagos Islands showing variation in beak form due to adaptation to specific niches. They may have originated from an ancestor who lived on the mainland 1 to 3 million years ago.

Darwin returned home in 1836 and devoted the remaining years of his life to the study of the data he had collected. It was fortunate that Darwin did not have to work for a living, because he had the time to think about the issues of variation, formation of new species, adaptation, and extinction and to read sources in geology, paleontology, taxonomy, demography, and evolutionary biology.

GEOLOGY: JAMES HUTTON AND CHARLES LYELL

Even in the Age of Reason, few European scholars would challenge the Genesis story of the creation of the earth. In addition, the Church supported the notion of a young Earth—October 23, 4004 B.C.—based on the calculations of the Irish Archbishop James Ussher. Reverend John Lightfoot, working independently, confirmed Ussher's creation date (Campbell et al. 2006). But some scholars strongly disagreed with Ussher's conclusion. One of these scholars was a naturalist, the Comte de Buffon (1707–1788).

In his book entitled *Histoire Naturelle* (1749), Buffon suggested that the earth had been shaped over millions of years by natural processes, such as wind and soil erosion, volcanic activity, and earthquakes

(Campbell et al. 2006). He was soon forced by the Catholic Church to recant his "heretical" notions of deep time or the ancient formation of the earth.

With advances in geology and the loosening grip of the Church over secular life, a new generation of scholars began to accept the theory of deep time. The layers and layers of sedimentary deposits seen on the sides of cliffs or on the sides of excavated trenches were too compelling to dismiss. James Hutton (1726–1797), a member of this new generation, described the earth as a decaying and self-renewing machine (Campbell et al. 2006; Larsen 2008). In his book entitled *Theory of the Earth* (1795), he wrote that terrestrial products such as soils and rocks decay through erosion and these products are carried into the sea. Subsequently, these erosion products are laid down as horizontal marine deposits, which eventually solidify into rocks. And later, these rocks are uplifted as continents by natural forces. Then the process repeats itself. But this was all speculation, albeit strong speculation. The proof for Hutton's hypothesis would come thirty years later.

In the meantime, the French paleontologist and comparative anatomist Georges Cuvier (1769–1832) was discovering the bones of long-extinct primates, dinosaurs, and mammoths in the rocks near Paris. He was able to show that at least 90 of the species he had recovered had become extinct (Campbell et al. 2006). Cuvier could not understand how such a massive loss of life could have occurred. In addition, he wondered why the bones of these prehistoric animals, particularly those of the dinosaurs, were so different from the bones of animals of the 18th and 19th centuries. "What destructive forces led to extinction of species, and what restorative forces controlled their replacement by later life forms?" (Campbell et al. 2006, 7).

Cuvier proposed his idea of **catastrophism** where global warming and cooling over time led to uplifting of the seafloor and caused sea levels to rise. Subsequently, there was a world-wide flood or "catastrophe" that wiped out many animal species over large areas. He believed that there were several of these sudden catastrophes. After the catastrophe, according to Cuvier, devastated areas were repopulated by different species of animals from other areas. And, to help with the repopulation, God created a few new species (Campbell et al. 2006; Larsen 2008). Cuvier never accepted the idea that the environment itself or needs of an animal can stimulate change. In essence, he rejected Jean Baptiste Lamarck's (1744–1829) evolutionary hypotheses. Nonetheless, Cuvier provided additional evidence for the ancient formation of the earth, and he is associated with the idea of extinction.

Building on the earlier work of Hutton, Sir Charles Lyell (1797–1875) proposed that the earth's geological history could be explained by such processes as erosion, volcanic eruptions, earthquakes, and floods, to name a few, that occur today. He concluded that these forces are *uniform*. His three-volume manuscript entitled *Principles of Geology*, published between 1830 and 1833, detailed the evidence for the ancient formation of the earth. William Whewell (1794–1866), a reviewer of *Principles of Geology*, coined the term *uniformitarianism* for Lyell's theory (Boaz and Almquist 2002). **Uniformitarianism** is the principle that processes that occur in the past are still occurring today. Specifically, Lyell proposed the following:

1. *Uniformity of law*: Through space and time, natural laws remain constant.
2. *Uniformity of process*: When possible, past phenomena should be explained as the results of processes now in operation.
3. *Uniformity of rate*: Change usually occurs slowly and steadily (unlike Cuvier's series of sudden catastrophic events).
4. *Uniformity of state*: Change, although continuous, is nondirectional and nonprogressive (Campbell et al. 2006; Gould 1987).

Lyell became one of the most influential geologists of the 19th century as he devoted most of his time writing about uniformitarianism and its implications for explaining Earth's history. And his calculations of how long mountain-building took provided the proof that Earth was millions of years old (Larsen 2008). Ironically, Lyell provided the time for large-scale evolution to occur through his empirical geological work but he rejected evolution as the cause for the diversity of species. So, like Cuvier, he would reject Lamarck's evolutionary hypotheses of the early 19th century.

The French naturalist Jean Baptiste Lamarck (1744–1829) was the first scientist who tried to explain how evolution worked. He believed, correctly, that there was a dynamic relationship between the organism and its environment. In his book entitled *Philosophie Zoologique* (1809), Lamarck discussed his major propositions. The first was the Law of **Use and Disuse**. According to Lamarck, a trait that is used frequently is enhanced over time. For instance, a monkey that leaps from tree to tree would develop stronger legs over time. In contrast, these same legs would grow weaker and wither away if the monkey remains sedentary. The second, more controversial law is the **Inheritance of Acquired Characteristics**. According to Lamarck, an organism will acquire a trait during its lifetime because of "need" and pass this newly acquired trait (at the specific stage of its development) to its offspring. Lamarck used this hypothesis to explain how the giraffe got its long neck. The original short-necked ancestor needed to reach the fruits on the top branches of tall trees. Consequently, this ancestor kept stretching its neck until the neck got progressively longer (Figure 2.4).

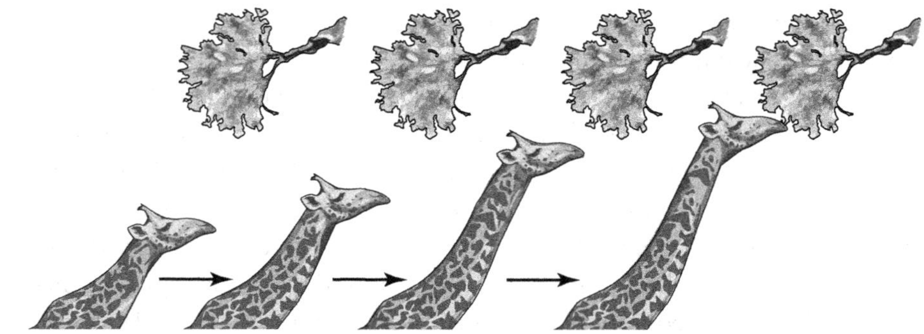

Figure 2.4. The inheritance of acquired characteristics: Lamarck's explanation of how the giraffe got its long neck.

For much of the 19th and early 20th centuries, Lamarck's hypothesis was taken seriously at the expense of Darwinian evolution. But with the rediscovery of Mendel's (see Chapter 3) work in 1900 and its synthesis with Darwinian evolution forming the Modern Synthesis and the discovery of DNA, Lamarckian evolution was proven wrong. Traits cannot be acquired during an organism's lifetime and then passed to its offspring. For a trait to be passed on to the offspring, it must be coded in the genes (egg and sperm).

What Lamarck did not understand was how environmental competition, rather than an organism's need, can drive evolutionary change. Nonetheless, a British economist and demographer, Thomas Malthus (1766–1834), made the connection between competition and survival. Malthus wrote *An Essay on the Principle of Population* (1798) where he observed that human populations increase geometrically whereas food supply increases only arithmetically. Specifically, he stated that there is not enough food for everyone born, so population is limited by food supply. Those individuals who can compete successfully will be able to feed their offspring, and their offspring will survive to adulthood. You will see later how Darwin will apply this demographic model to his theory of natural selection.

DARWIN, WALLACE, AND THE THEORY OF NATURAL SELECTION

Despite the fact that only "original" research is the criterion used for obtaining a Ph.D. or getting your work published in a scholarly journal, no work is really original. If you search deep enough, you will find that the so-called original work is based on the work of previous scholars. This was the same for Darwin. The context for Darwin, as mentioned earlier, was geology, paleontology, taxonomy, demography, and evolutionary biology. Geology in the form of Lyell's work gave Darwin the *time* he needed for evolution to work. According to Darwin, evolution needs time for gradual, slow change and the formation of new species. From paleontology or Cuvier's work, Darwin took away the idea of extinction. He believed that the fate of all species is extinction. Studying Linnaeus's taxonomy, Darwin noted the similarities in skeletal traits between apes and humans. He concluded that humans must be related in some way to these apes. Furthermore, reading Malthus's *Essay*, Darwin concluded that in an environment of competition, those individuals with favorable traits will survive and reproduce and those with unfavorable traits will die. Finally, he accepted Lamarck's conclusion that there was a dynamic relationship between the organism and its environment. However, Darwin believed that the environment itself stimulated change and not the organism's needs. More importantly, he believed that variation in the population was necessary for change and survival (see Figure 2.5). Consequently, Darwin's theory of evolution through natural selection is as follows: Individuals have more offspring than can survive. Because of this, there will be ***struggle for existence***. Those individuals with favorable (adaptive) traits will survive and reproduce; those with unfavorable (nonadaptive) traits will die—***variation in fitness***. Those individuals who survive and reproduce will pass their traits on to the offspring—***inheritance of variation***. There will be a higher frequency of these traits in future generations. Darwin got the idea for using the words "natural selection" from animal breeders who were trying to obtain certain traits artificially by breeding male and female animals carrying those desired traits.

After many years of thought and observation, Darwin published his revolutionary book *On the Origin of Species by Means of Natural Selection Or the Preservation of Favoured Races in the Struggle for Life* (1859) (Campbell et al. 2006). Darwin was hesitant in submitting his controversial work for publication. But he was encouraged to publish when Alfred Russel Wallace (1823–1913) (Figure 2.6) sent him a copy of his paper. Wallace, working independently, had come to similar conclusions. Wallace studied species origins for five years in the interior of Brazil, and then he relocated to Indonesia where he studied the natural history of the islands in the Malay Archipelago (Boaz and Almquist 2002). While Darwin is generally recognized as the discoverer of the theory of natural selection, Wallace also gets some credit for contributing substantially to evolutionary theory (Larsen 2008).

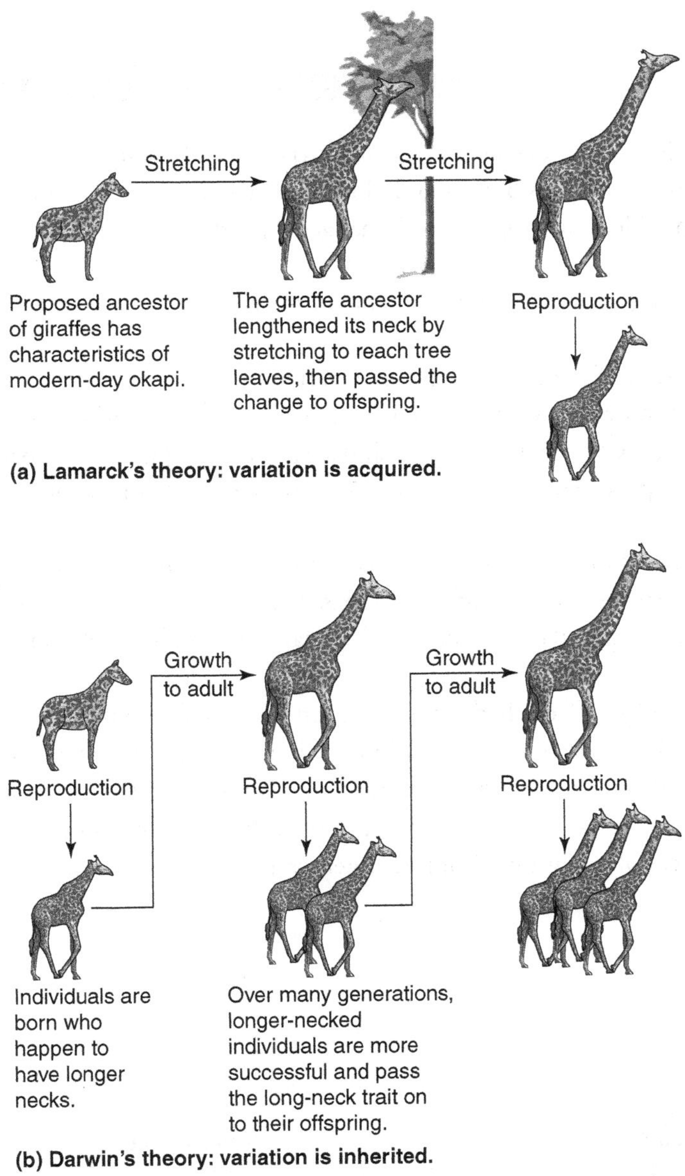

Figure 2.5. A comparison of (a) Lamarck's and (b) Darwin's evolutionary theories (from Raven and Johnson 2002).

Figure 2.6. Alfred Russel Wallace, co-founder of the theory of natural selection.

Some important principles of evolution that were developed in the mid-20th century are as follows:

1. No trait is "good" or "bad." It depends on the specific environment.
2. Biological traits can change in different directions. For example, there was a steady *increase* in human cranial capacity from 1.0 million years to 100,000 years. Simultaneous, there was a *reduction* in the jaw and teeth.
3. Evolution does not occur in a vacuum. The environment provides the context.
4. Variation is critical in the evolutionary process.

Darwin knew much about variation and the fact that it is passed from the parent to offspring. But he did not know the sources of variation within each generation and the mechanism of inheritance. Interestingly, Gregor Mendel (1822–1884), an Augustinian monk living in a monastery in what is now Brno, Czech Republic, was addressing these problems. We can assume that due to the slow information system of the 19th century and the language barrier, Darwin was unaware of Mendel's work.

Summary of Key Ideas

1. Before Darwin proposed his theory of evolution, there was no context for variation and change because the worldview was stasis or the "fixity of species." God was perfect and God created a perfect Earth. Since religion was prominent in society at this time, ideas of variation and change implied that God was not perfect. This was heresy.
2. Aristotle's *scalae naturale* or Great Chain of Being was the dominant theme in the Medieval, Renaissance, and Enlightenment periods.
3. During the Enlightenment, the ancient formation of the earth (deep time), variation, and the dynamic relationship between the organism and the environment were established by scientists in five scientific disciplines: geology (Hutton and Lyell), paleontology (Cuvier), taxonomy (Linnaeus), demography (Malthus), and evolutionary biology (Lamarck). They also provided the context for Darwin to generate his theory of evolution.
4. Darwin's theory of evolution through natural selection states that in an environment of competition, those individuals with favorable (adaptive) traits will survive and reproduce and pass their traits to the offspring. As time goes on, these traits will increase in the population. Those individuals with unfavorable traits (nonadaptive) will die.
5. Important principles of evolution: (a) No trait is "good" or "bad." It depends on the specific environment. (b) Biological traits can change in different directions; for example, human brain size got larger while the teeth and jaw got smaller. Again, it depends on the environment. (c) Evolution does not occur in a vacuum—the environment provides the context. (d) Variation is critical in the evolutionary process.
6. Darwin did not know the following: (a) sources of variation within each generation; and (b) the mechanism of inheritance.

Key Terms

Adaptation
Catastrophism
Enlightenment
Fixity of species
Galapagos Islands
Great Chain of Being
Heliocentric
Inheritance of Acquired Characteristics
Medieval
On the Origin of Species
Renaissance
Stasis
Uniformitarianism
Use and Disuse

Further Readings

Browne, J. 1995. *Charles Darwin: Voyaging.* New York: Knopf.

Gould, S. J. 1992. *Ever since Darwin: Reflections on natural history.* New York: Norton.

Moorehead, A. 1969. *Darwin and the Beagle.* New York: Penguin Books.

Repcheck, J. 2003. *The man who found time: James Hutton and the discovery of Earth's antiquity.* Cambridge, MA: Perseus Publishing.

INTERNET RESOURCES

Pre-Darwinian Theories
http://anthro.palomar.edu/evolve/evolve_1.htm
This is the Anthropology Department at Palomar College's Web site on pre-Darwinian evolutionary hypotheses.

UCMP Exhibition Halls: Evolution
http://www.ucmp.berkeley.edu/history/evolution.html
This Web site provides biographical sketches of several scientists who contributed to theories about evolution and the ancient formation of the earth (deep time).

Victorian Science: An Overview
http://www.victorianweb.org/science/sciov.html
This Web site provides background information about Charles Darwin and other 19th-century scientists.

Chapter 2: Evolutionary Theory

Study guide: By knowing the material below you will be well prepared for the exam. Don't forget to read the chapter and study your notes (this includes the chapter worksheets).

1. Be able to discuss *blending inheritance*
2. Be able to name the scientists whose combine effort helped to learn the structure of DNA
3. Be able to name the first scientist who tried to explain evolutionary change
4. What was the context for Darwin to generate his theory of evolutionary change (natural selection)? In other words, his theory drew from many scientific disciplines. Name and discuss.
5. Why is Alfred Russell Wallace significant?
6. Why is Carolus Linnaeus significant?
7. *Uniformitarianism* is associated with a British geologist. Be able to name him and explain uniformitarianism
8. The earth is a 'decaying and self-renewing machine.' Explain this proposal. Who is the naturalist responsible for this proposal?
9. Explain the "inheritance of acquired characteristics?" Who was responsible for this hypothesis?
10. What did the British demographer Malthus say about population and food supply?
11. Which scientist coined the term *Catastrophism*? Be able to explain *Catastrophism*
12. The controversial notion of extinction is associated with the following scientist named________.
13. Be able to define mutation. It is the only source of______________.
14. What was the worldview about human origins before modern times?
15. Be able to explain natural selection.

16. According to Darwin's theory of natural selection, disadvantageous traits will disappear over time in a population. True/False?
17. When Darwin coined the term "natural selection" he was making an analogy to artificial selection. True/False?
18. Variation in a population is the key to Darwin's idea of natural selection. True/False?
19. What did Darwin not know about heredity?
20. "Fitness," in an evolutionary sense, refers to an individual's_______________.

CHAPTER 3

Genetics and Heredity

After he articulated his theory of evolution through **natural selection**, Darwin addressed the problem of heredity. Specifically, he wanted to know *how* traits being selected for (or against) were passed from parent to offspring (Larsen 2008). Many people could readily observe that children resembled their parents. For example, the child has her father's eye color and her mother's nose form. Darwin and his contemporaries believed that this phenomenon was caused by a sort of **blending inheritance** mechanism. From personal observations and reading, Darwin hypothesized that each body part contained invisible particles called **gemmules** (Larsen 2008). Furthermore, he believed that the gemmules for each body part resided in the reproductive organs. During fertilization, the mother's and father's gemmules come together or "blend" to form characteristics observed in the offspring. But the science of genetics, based on **Mendel**'s work, will show that these invisible particles or "genes" do not blend. They are inherited independently of each other.

In his early years, Mendel (Figure 3.1) showed great aptitude for math and science. In addition, he learned much about plants and horticulture from his father, who was a small-scale farmer and fruit grower (Campbell et al. 2006). Mendel enrolled in the University of Vienna to study natural science and received an excellent education. But he could not complete his university education because of the onset of anxiety before major examinations: he would get physically ill (Boyd and Silk 2006). Mendel left the University of Vienna and entered the Augustinian monastery in the city of Brno, where he continued botanical studies on inheritance and variation.

Figure 3.1. Gregor Mendel, whose experiments with the garden pea plant laid foundations for genetics.

From 1859 to 1865, Mendel crossbred different varieties of garden pea plants. He was more successful, than his predecessors in elucidating the mechanism of inheritance because he kept careful count of the progeny in each generation and wrote everything down (Figure 3.2). In addition, he chose the ideal experimental subject—the garden pea (*Pisum sativum*). The garden pea is easy to grow, has a short generation time, contains both male and female sex organs, and has a large number of **true-breeding** (pure) varieties. Now let us study Mendel's experimental design and results using one variety of the pea plant, flower color.

Before Mendel performed cross-fertilization, he allowed the pea plants of a given variety to produce by self-fertilization. This initial procedure was critical because he wanted to make sure that the varieties were true-breeding and not the result of natural cross-breeding. In short, he did not want hybridization complicating his experiment at this stage. After he was satisfied that the varieties were true-breeding, he performed his **cross-fertilization**. He removed the male parts (anthers that contain the pollen) from a purple flower, and then transferred pollen from a white flower to the female parts (stigma) of the purple flower (Figure 3.3; the white flower is on the left and the purple is on the right). In the F_1 (first filial) generation, Mendel saw all purple flowers (phenotype; Figure 3.3). At this point, he had a basic suspicion that the purple color was the dominant trait and the white color was recessive. He allowed the F_1 generation or hybrids to self-fertilize. In the F_2 generation, Mendel saw three purple and one white (Figure 3.4). Mendel's suspicions were confirmed: the white color was the recessive trait. It had skipped a generation and reappeared in the subsequent generation. While the **phenotype** or visible characteristics of the trait, in this case, color, could be clearly seen, Mendel did not have the technology to see the **genotype** or underlying genes. But he realized that a *discrete* physical unit was responsible for each trait, and this physical unit was inherited (Larsen 2008). Today, we call this physical unit a **gene**. Now let us go back and study the results of Mendel's experiment, making note of the underlying genes. If we use the genotype 'AA' to represent the purple flower and 'aa' to represent the white flower, in the F_1 generation the genotypes were all 'Aa' or **heterozygote** dominant (Figure 3.5a). The **dominant** allele (alternate form of a gene) 'A' (purple) is always expressed; it masks the recessive allele. For a **recessive** trait to be expressed, the offspring must inherit the alleles coding for the trait from both parents. A **Punnett square** can be used to show the percentages of different genotypes after respective matings (see Figure 3.5).

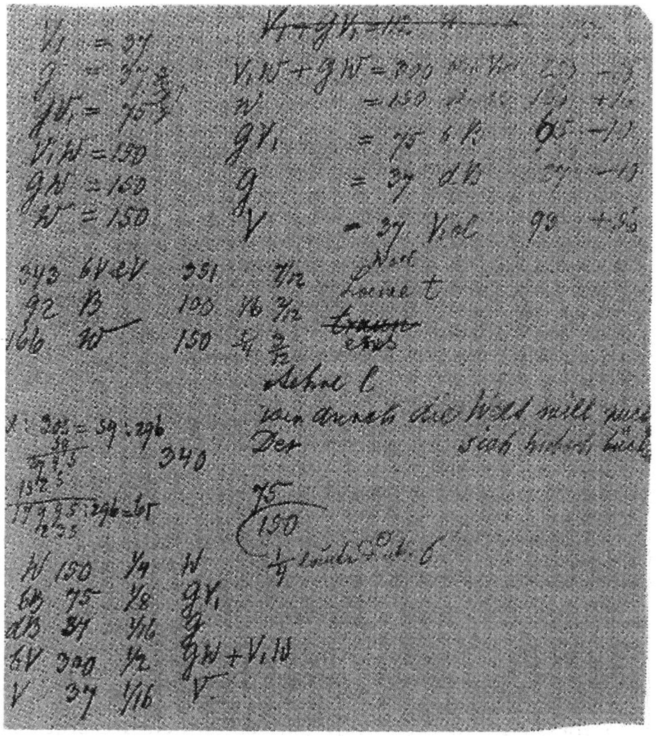

Figure 3.2. Mendel kept careful count of the number and characteristics of the progeny each generation.

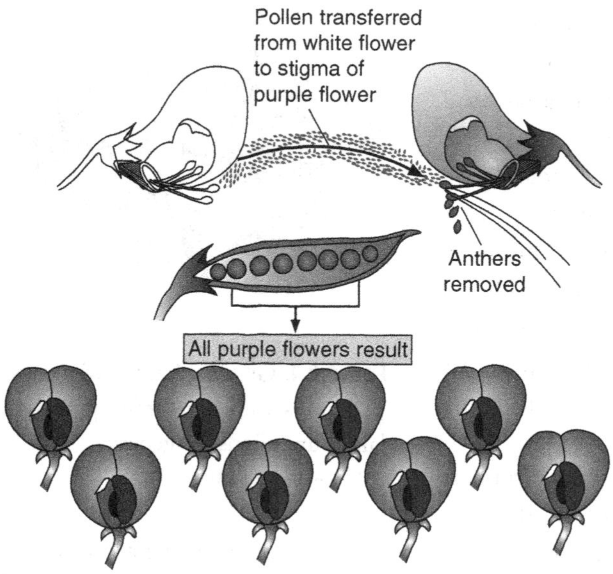

Figure 3.3. Mendel's F_1 generation was all purple after crossing a white flower with a purple flower.

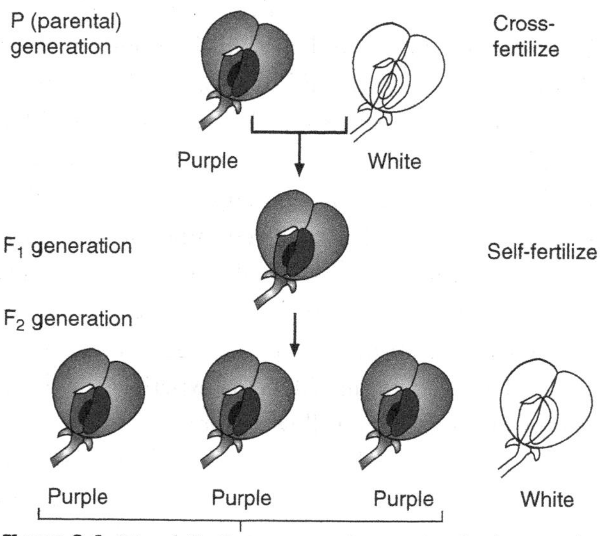

Figure 3.4. Mendel's F_2 generation comprised of 3 purple to 1 white. The white flower skipped one generation and reappeared in another.

The **hybrids** or heterozygote dominants 'Aa' were allowed to self-fertilize. In the F_2 generation, the genotypes were one 'AA' or **homozygous** dominant; two 'Aa' (heterozygote dominant); and one 'aa' or homozygous recessive (Figure 3.5b). The **genotypic ratio** is 1(25%):2(50%):1(25%); the phenotypic (or **Mendelian**) **ratio** is 3:1.

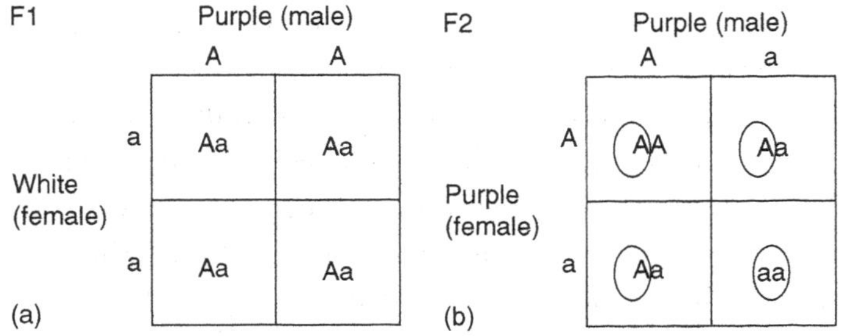

Figure 3.5a,b. The genotypes in the F_1 generation are all heterozygote dominant. In the F_2 generation the genotypes are homozygote dominant (25%); heterozygote dominant (50%); and homozygous recessive (25%).

Mendel eventually crossed a pea with two dominant traits (round and yellow) with a pea having two recessive traits (wrinkled and green), producing many combinations of the four traits with a 9:3:3:1 ratio (Figure 3.6). He was able to formulate two principles to explain his experimental results: (1) every trait has two discrete units, one inherited from the mother and one from the father. Modern scientists call this the **law of segregation**; and (2) the inheritance of one trait (or discrete unit) does not affect the inheritance of other traits (or discrete units). Modern scientists call this the **law of independent assortment**. Interestingly, these principles were confirmed by research on chromosomes and cell division (meiosis) in 1902.

Mendel's paper was read to the Brno Society for the Study of Natural Science in 1865 and published in that society's proceedings in 1866 (Boyd and Silk 2006; Campbell et al. 2006; Larsen 2008). One could speculate that the obscure nature of this local scientific journal doomed Mendel's work: no one could grasp the importance of this work on heredity. In fact, Mendel sent a copy of his paper to Karl Nägeli (1817–1891), a very prominent botanist, who dismissed it as unimportant. After 1866, Mendel devoted the rest of his life to monastic duties. In effect, Mendel's work would be forgotten for more than 30 years.

In 1896, the Dutch botanist Hugo de Vries (1848–1935) unknowingly repeated Mendel's experiment using the primrose. From 1896 to 1900, he conducted several experiments before he was confident that his results could not be challenged. But when de Vries was ready to publish his results, a colleague sent him a copy of Mendel's 1866 paper. Mendel's work was finally

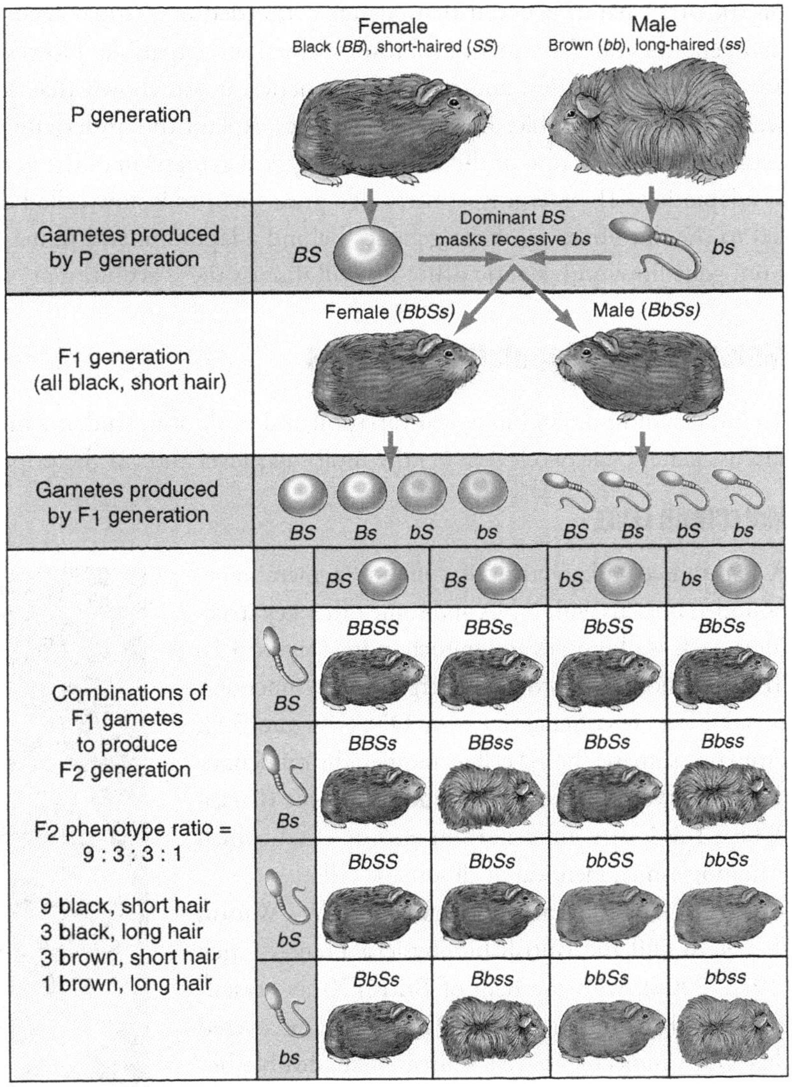

Figure 3.6. Mendel's dihybrid cross using round versus wrinkled and yellow versus green seeds. The ratio of the four possible combinations of phenotypes is 9:3:3:1. **Note: Mendel never used guinea pigs; nonetheless, the phenotypical ratios for simple inheritance of discrete traits are identical for many species.**

rediscovered. Two other researchers—the German botanist Carl Erich Correns (1864–1933) and the Austrian botanist Erich Tschermak von Seysenegg (1871–1962)—working independently also rediscovered Mendel's paper and replicated his results (Boyd and Silk 2006). In essence, the science of genetics began in 1900.

By 1900, careful embryological research had generated information on the **cell**, preliminary information about cell division, and thread-like cellular structures called chromosomes. **Chromosomes** are strands of DNA in the nucleus of cells containing thousands of genes. In 1902, Walter Sutton (1877–1916) of Columbia University understood why the discrete physical units, identified as genes by the American geneticist Thomas Hunt Morgan (1866–1945), were inherited in equal proportions by the offspring from mother and father. Since genes are on chromosomes, each somatic (body) cell receives a copy of both maternal and paternal chromosomes during mitosis (mitosis is discussed later in the chapter). Also, traits segregate independently of each other because the genes that control them are on different chromosomes. In other words, each chromosome strand segregates independently during cell division.

As great strides in embryological research continued, a great chasm had developed between proponents of Mendelian genetics and traditional Darwinians. Specifically, Mendelian genetics proved that inheritance was discrete: pea shape can only be round or wrinkled, not some roundness and some wrinkledness. The Darwinians, on the other hand, believed that change proceeded by gradual accumulation of small changes. In short, Mendelian genetics and Darwinian evolution seemed incompatible. However, in the 1930s, 1940s, and 1950s with the advances in population and molecular genetics, it was shown that Mendelian genetics and Darwinian evolution were indeed compatible: Mendelian genetics explain the underlying genetic changes while Darwinian evolution explains the interaction of the traits (the physical expression of the genes) with the environment commonly known as adaptation. The merging of these two great theories is now called the **modern synthesis**. The modern synthesis led to the illumination of the genetics behind adaptation and non-Darwinian evolutionary forces, such as mutation, gene flow, and genetic drift. We will discuss these evolutionary forces in Chapter 4.

Molecular and Population Genetics

To understand human biological variation and evolution, students must understand basic genetics. We will briefly discuss genetics at two levels: (1) the molecular level and (2) the population level (or **microevolution**).

MOLECULAR LEVEL

A generalized eukaryotic cell contains a membrane-bounded nucleus with DNA and some other key structures, such as ribosomes and mitochondria (Figure 3.7). The mitochondrion provides energy for cell functions. In addition, it contains its own DNA (Figure 3.8). From one somatic (body) cell to another, the mitochondria are different. It is said to be heteroplasmic (Larsen 2008). This is very different from nuclear DNA, which is homoplasmic, identical in all somatic cells.

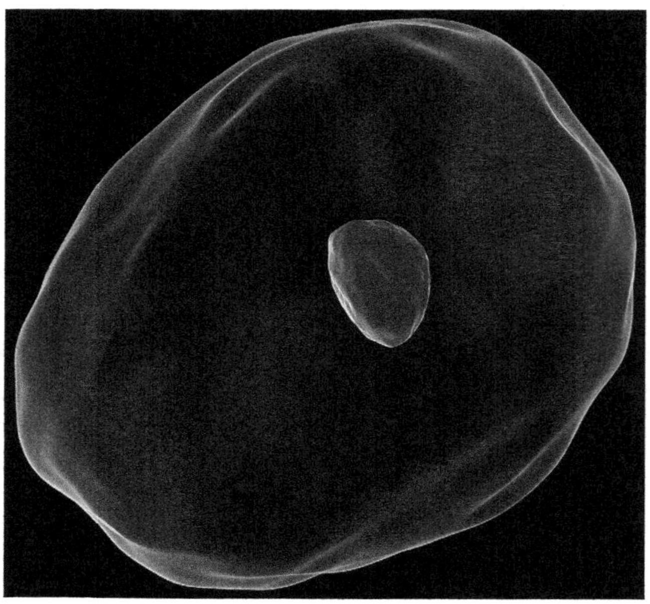

Figure 3.7. A eukaryotic cell.

Image © Sebastian Kaulitzki, 2010. Used under license from Shutterstock, Inc.

In 1953, the American geneticist James Watson (b. 1928) and the British biophysicist Francis Crick (1916–2004), with the help of British X-ray crystallographer Rosalind Franklin (1920–1958), discovered that DNA molecules have a staircase-like, **double helix** structure (Figure 3.9). **DNA** (a simplified version of the word *deoxyribonucleic acid*) contains a phosphate, sugar, and four nitrogenous bases: adenine (A), thymine (T), guanine (G), and cytosine (C). Adenine

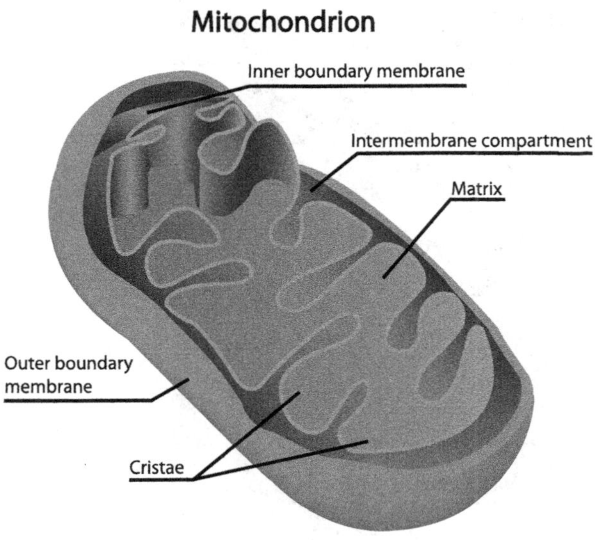

Figure 3.8.

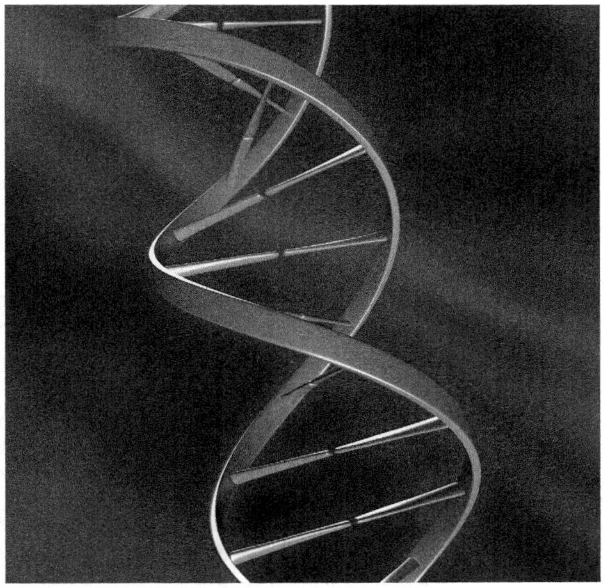

Figure 3.9. The DNA molecule (double helix)

is always bonded to thymine and guanine is always bonded with cytosine. The binding of a phosphate, sugar, and base is called a nucleotide. Nucleotides form the rails of the "staircase" while the bases form the rungs (Figure 3.10). The bond that connects the bases is very weak so that DNA can easily replicate itself for important functions, such as mitosis, meiosis, and protein synthesis.

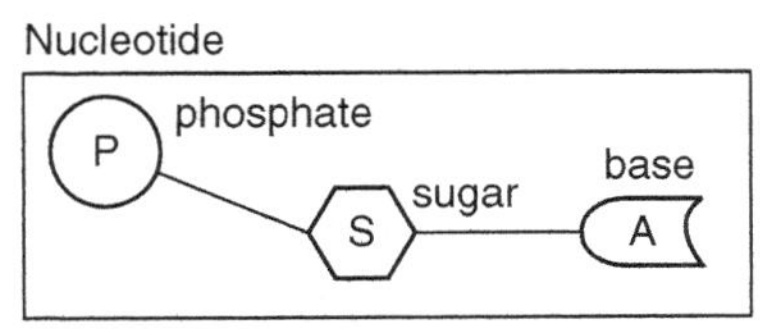

Figure 3.10. A nucleotide consists of a phosphate, sugar, and one of the four nitrogenous bases.

CHROMOSOMES

As mentioned earlier in this chapter, chromosomes are strands of DNA in the nucleus of cells containing thousands of genes (Figure 3.11). Humans have 46 chromosomes (or 23 pairs) (Figure 3.12). In other words, our somatic cells, that is, heart, liver, skin, brain, to name a few, have the full complement (**diploid**) of chromosomes. But our sex cells (egg and sperm) only have 23 chromosomes or half (**haploid**) of the total number of chromosomes (Figure 3.13). In summary, there are 22 pairs of **autosomal chromosomes** and one pair of sex chromosomes: males designated by the letters XY and females designated by the letters XX (Figure 3.14a,b). It is important that each cell gets the correct number of chromosomes in mitosis and meiosis because mistakes—cells receiving more or less than the correct number of chromosomes—can lead to debilitating mutations or death.

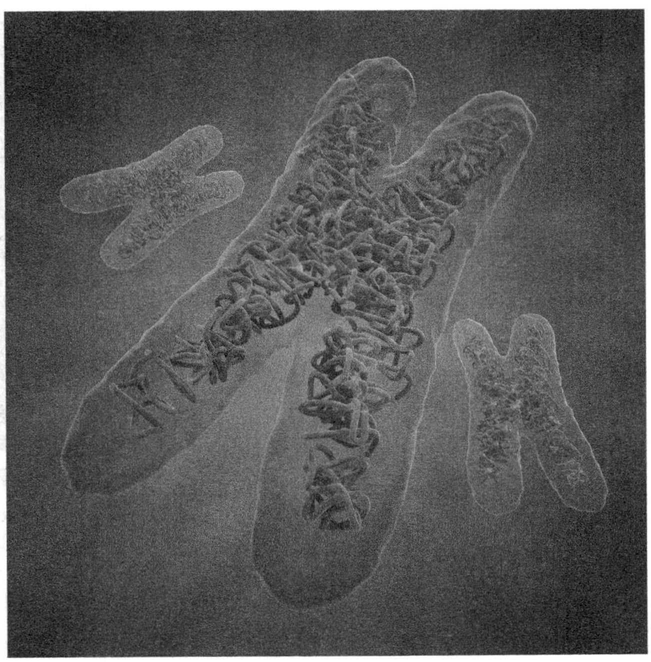

Figure 3.11. Several chromosome pairs.

MITOSIS AND MEIOSIS

Mitosis is the production of somatic cells (Figure 3.15a). New cells are produced during growth and

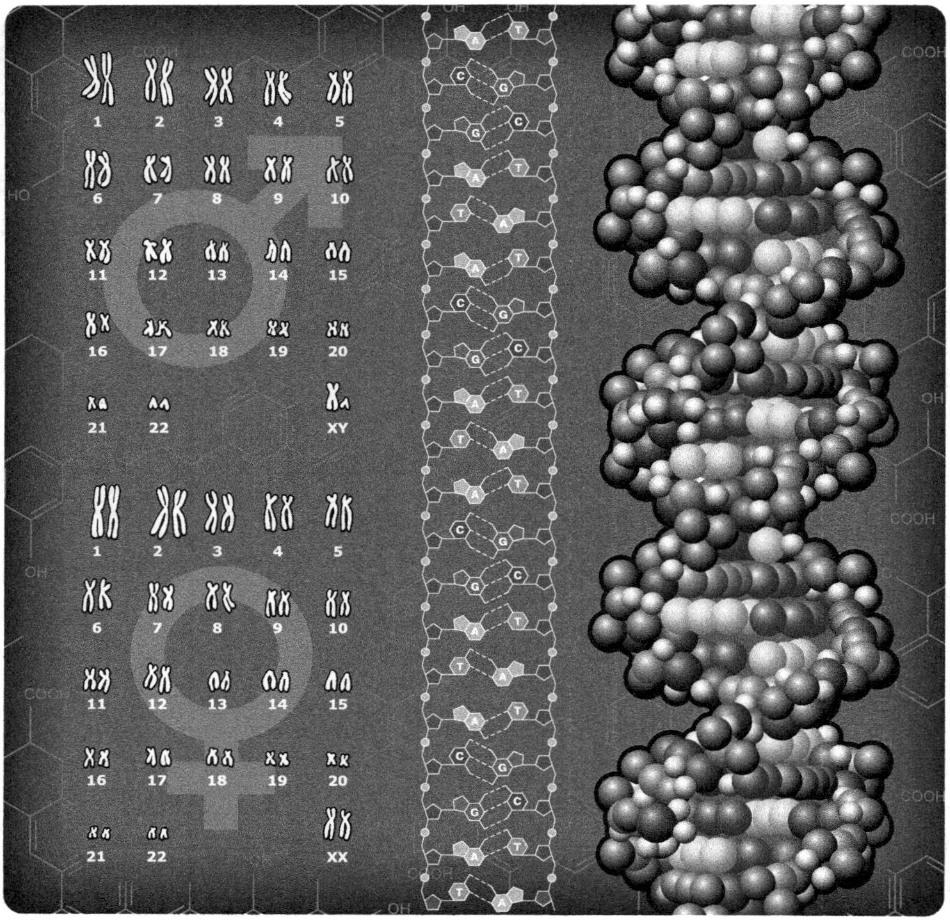

Figure 3.12. The full complement of chromosomes in a human being (the human genome).
Image © Ykh, 2010. Used under license from Shutterstock, Inc.

development and injury. If you shed or cut your skin, within a few days or weeks mitosis will generate new skin. **Meiosis**, on the other hand, is sex cell production (Figure 3.15b).

It is a special form of cell division where genetic variation is produced. Specifically, there are two stages in meiosis. In the first stage, **crossing over** occurs when chromosomes come together and exchange strands (Figure 3.16). Since genes reside on chromosomes, **recombination** occurs, producing variation. In the second stage, meiosis proceeds similar to mitosis but producing cells with half the number of chromosomes.

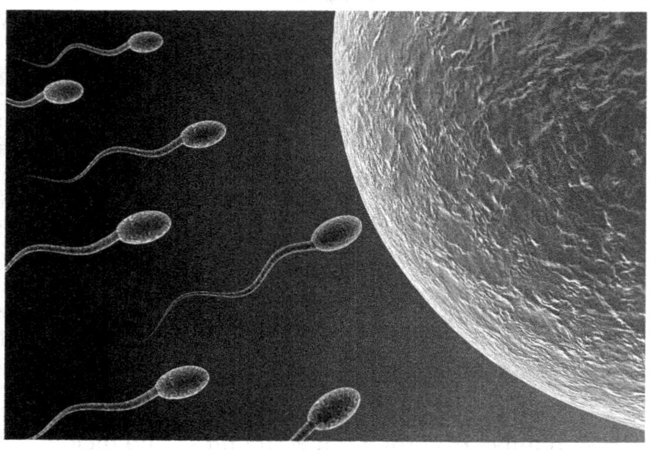

Figure 3.13. The sperm and the egg only have 23 chromosomes or half of the total number.
Image © Cre8tive Images, 2010. Used under license from Shutterstock, Inc.

PROTEIN SYNTHESIS

Genes translate into proteins. DNA must replicate itself in order to produce proteins. But DNA cannot leave the nucleus of the cell. Consequently, it transcribes (**transcription**) a copy that can leave, which is called messenger RNA (mRNA). Messenger RNA, although a copy of DNA, has a few important differences: it is a single strand; it has the base uracil instead of thymine—but still binds with adenine; and it can leave the nucleus. After

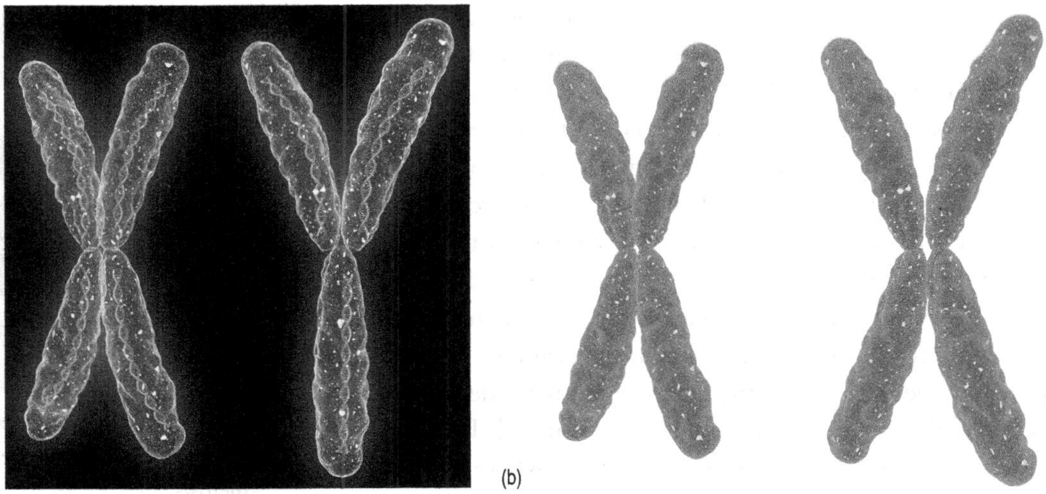

(a) (b)

Figure 3.14a,b. (a) human male; (b) human female

Image © Sebastian Kaulitzki, 2010. Used under license from Shutterstock, Inc.

Image © Sebastian Kaulitzki, 2010. Used under license from Shutterstock, Inc.

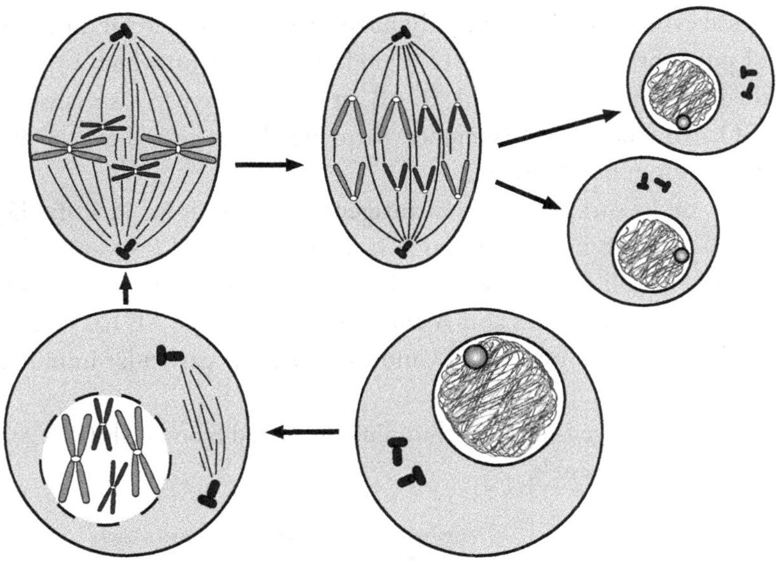

(a)

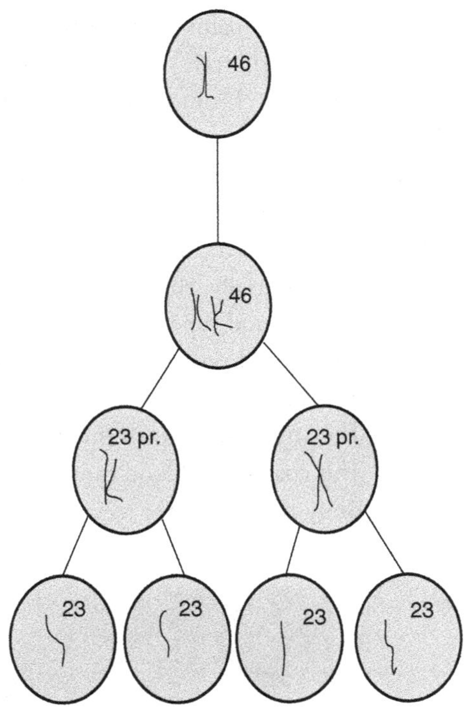

(b)

Figure 3.15a,b. (a) Mitosis: somatic cell production; (b) Meiosis: sex cell production.

Image © Dreamy Girl, 2010. Used under license from Shutterstock, Inc.

the mRNA leaves the nucleus, it reaches the ribosome and binds to it. Then the genetic message is translated (**translation**). Genes translate into protein, that is, structural proteins like collagen in bones and keratin in skin or gas transport proteins like hemoglobin. In addition, there are regulatory proteins that turn other genes on and off, for example, in growth and development. These **regulatory genes** might be the reason why chimpanzees and humans are very physically different despite the 98 percent genetic similarity: there are drastically different growth and development trajectories in chimpanzee (and gorillas) and human infants as they leave infancy (Figures 3.17a,b and 3.18a,b).

Regulatory genes are implicated in the varying frequencies of **lactose intolerance** in human populations. In other words, the enzyme (special protein) lactase, which breaks down lactose (milk sugar), is "turned off" in some populations. As a result, these populations cannot digest dairy products. Dairy products ferment in the gut and causes serious diarrhea (in lactose-intolerant individuals). It is interesting that all human infants are lactose tolerant simply because these infants need milk. Consequently, the enzyme lactase is "turned on." So what is the reason for lactose intolerance in some populations after weaning? At this point we come back to a concept we discussed in Chapter 1, which is biocultural interaction: biology affects culture and culture affects biology. In this case, culture is affecting biology. In essence, the lactase gene stays active (after weaning) in cultures with a long tradition of using dairy products, that is, products from, for example, cows or goats, such as butter or cheese, because the lactase gene is needed in this environment where dairy products are frequently consumed. In contrast, the lactase gene is turned off (after weaning) in cultures where dairy products are not traditionally consumed, because biological evolution is not wasteful: what is not being used is turned off.

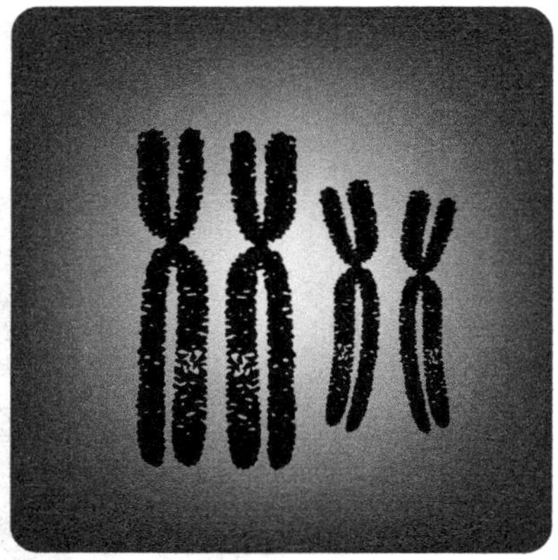

Figure 3.16. Crossing over of chromosome strands during the first stage of meiosis.

Image © attem, 2010. Used under license from Shutterstock, Inc.

GENETIC (BLOOD) POLYMORPHISMS

Genetic polymorphisms are important in biological evolution. The word **polymorphism** refers to several different alleles for a genetic trait. One good example of polymorphism is the ABO (blood group) system, which is positioned (**locus**) on chromosome 9. Each individual can have phenotypes A, B, AB, or O. Blood type O is recessive. The genotype combinations are AA AO, BB BO, AB, and OO. Remember, there are always two alleles in the genotype combinations because we inherit our genes from both parents.

Continuing, each red blood cell has proteins on its surface known as **antigens**. These antigens identify the cell as being A, B, AB, or O (no antigens) and will cause an immune reaction if the wrong blood is given to the individual. Essentially, the individual's immune system will produce **antibodies** to destroy this foreign material (Figure 3.19). Blood type A has anti-B antibodies; blood type B has anti-A antibodies; blood type AB has no antibodies; and blood type O has anti-A and anti-B antibodies (Table 3.1). Unfortunately, this particular immune response leads to several lethal effects:

1. The antibodies attach themselves to the foreign antigens causing clumping of the red blood cells (RBCs).
2. This coagulated blood cannot pass through the blood vessels.
3. Body tissues do not receive their supply of blood.

(a)

(b)

Figure 3.17a,b. Different growth and development trajectories in terms of facial projection in chimpanzees from (a) infancy (non-projecting) to (b) adulthood (projecting).

Image © Ronnie Howard, 2010. Used under license from Shutterstock, Inc.

Image © Holder Ehlers, 2010. Used under license from Shutterstock, Inc.

Table 3.1: ABO Blood Group System

Phenotypes	Genotypes	Antigens	Antibodies
A	AA AO	A	Anti-B
B	BB BO	B	Anti-A
AB	AB	A, B	None
O	OO	None	Anti-A, Anti-B

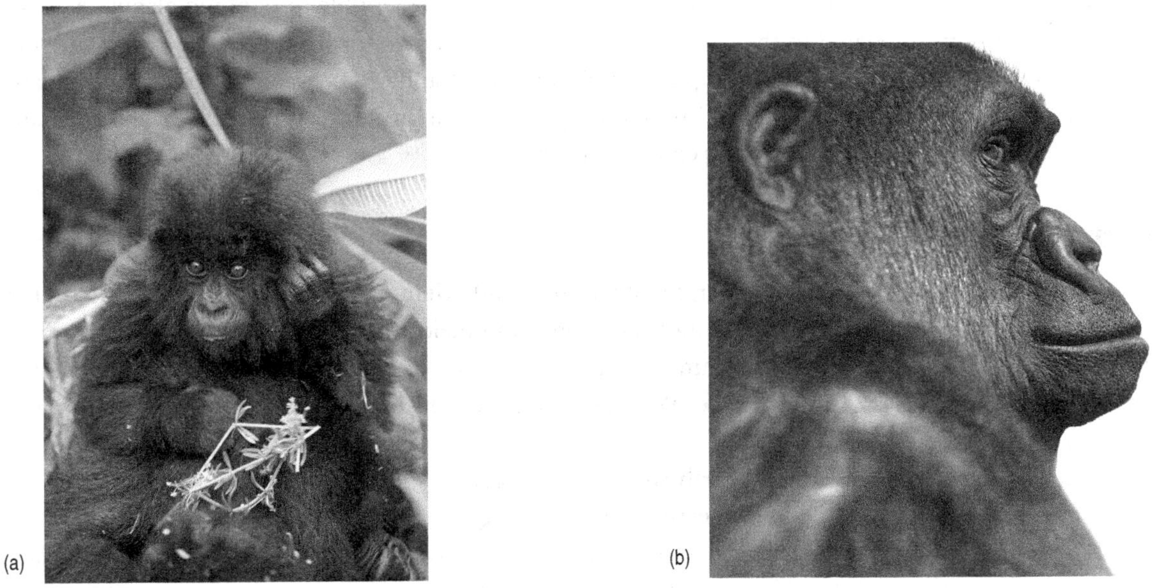

Figure 3.18a,b. Different growth and development trajectories in terms of facial projection in gorillas from (a) infancy (non-projecting) to (b) adulthood (projecting).

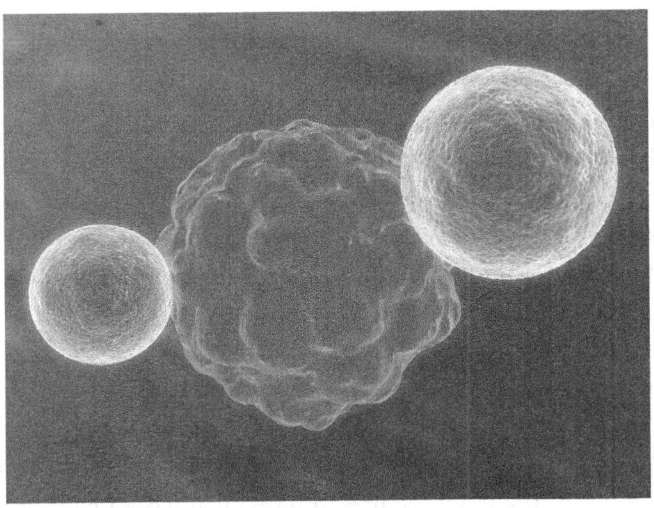

Figure 3.19. Antibodies attach themselves to the foreign antigen.

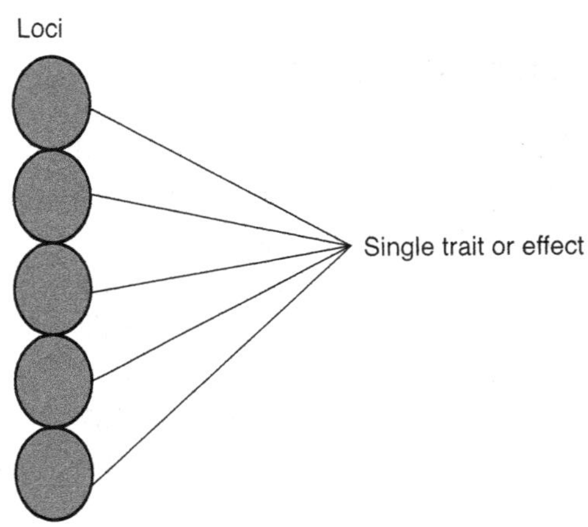

Figure 3.20. Polygenetic traits.

Fortunately, these lethal effects can be prevented because medical doctors have known about the biology behind red blood cell antigens and antibodies since the early 20th century.

ABO TYPES AND DISEASE HISTORY

The distribution of blood types around the world appears to be related to disease resistance and natural selection. The different antibodies of the ABO blood groups support the former statement because people with blood types B and O, for example, will have greater resistance to microorganisms that are biochemically similar to A molecules and, likewise, people with blood type A will have greater resistance to microorganisms that are biochemically similar to B molecules (Relethford 2005).

Based on past research (Molnar 1998; Relethford 2005), the ABO blood types appears to be related to infectious and noninfectious diseases. For instance, blood type A seems to be more susceptible to smallpox, type B seems to be more susceptible to infantile diarrhea, and type O seems to be more susceptible to bubonic plague. Asia has a high frequency of B blood type, related to the fact that this region was known for having smallpox and bubonic plague epidemics which selected against (killed) people with blood types A and O. As far as non-infectious diseases are concerned, it has been suggested that people with blood type O have a greater of getting stomach ulcers and people with type A have a greater chance of getting certain cancers.

POLYGENIC VARIATION AND PLEIOTROPHY

Mendelian genetics, discussed earlier in this chapter, involves simple inheritance. These traits are either present or absent and there is no environmental effect. In contrast, polygenic and pleiotrophic traits are considered complex traits because they can be measured on a continuum, they are affected by the environment, and they are not as predictable as Mendel's simple rules of **inheritance**. **Polygenic** refers to several loci coding for the same trait or effect (Figure 3.20).

Skin color, height, eye color, weight, and birth weight, for example, are traits coded by several loci and measured on a continuum. **Pleiotrophic** refers to a single gene coding for two or more traits or effects (Figure 3.21). Size and shape are correlated and coded by a single gene. Also, the sickle cell gene codes for a sickle-shaped red blood cell and growth/developmental abnormalities.

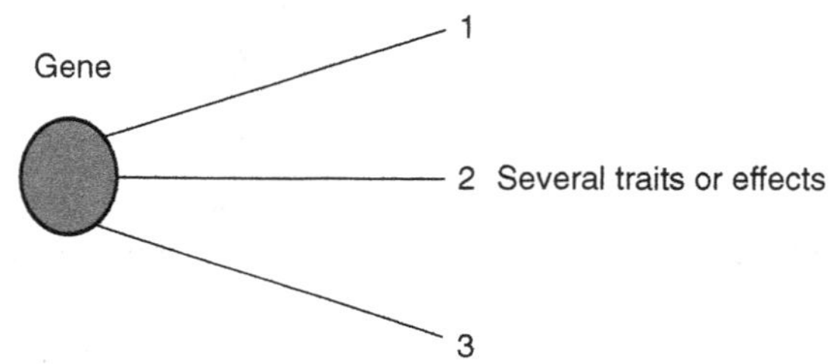

Figure 3.21. Pleiotrophic traits.

Population Genetics

Population genetics is the study of the relationship between allele and genotype frequencies in populations over a few generations (microevolution). Remember, in the late 19th century scientists did not know how traits were inherited despite the fact that Mendel had published his extensive work on heredity in 1866. The idea of blending inheritance was widely accepted, and most scientists believed that it would deplete variation, leaving nothing for selection to work on. Additionally, they believed that selection led to optimal forms and thus would eliminate variation.

In 1908, Godfrey Hardy (1877–1947), a British mathematician, and Wilhelm Weinberg (1862–1937), a German obstetrician, independently solved the problem of why variation persists. They proposed that if certain assumptions are met—(1) random mating is occurring in the population—any male has an equal opportunity to mate with any female and vice versa; (2) no mutation is occurring in the population—no new variation (new genes) in the population; (3) no natural selection is occurring in the population—no individual has a higher reproductive fitness than other individuals; (4) no gene flow is occurring in the population—no new genes are coming into the population (no immigration); and (5) no genetic drift—the population is very large—then genotype frequencies will remain constant from generation to generation. The genotypes are said to be in **Hardy-Weinberg equilibrium** or no evolution.

The Hardy-Weinberg equilibrium is a mathematical model ($p^2 + 2pq + q^2$) that shows the relationship between alleles and genotype frequencies. It is a hypothetical situation that can be used as a "barometer" or a "control" (*expected*) against which you can compare what is actually happening (*observed*) in the population. In short, you can find out whether evolution is occurring in the population. Let us look at the following example, keeping in mind that the goal of the model is to compare what is observed with what is expected based on the model. In this hypothetical population, there are 250 individuals with three genotypes: AA = 40, Aa = 120, aa = 90 (40 + 120 + 90 = 250 individuals):

AA	Aa	aa
40	120	90

*Step 1: Find the genotype frequencies: divide the genotypes by 250

AA	Aa	aa	
40/250	120/250	90/250	
0.16	**0.48**	**0.36**	**genotype frequencies (*observed*)**

*Step 2: Count the number of **A** and **a** alleles: total alleles in population = 250 × 2 = 500

	AA		**Aa**
	40		120
A alleles =	40 × 2	+	120 = 200
	Aa		**aa**
	120		90
a alleles =	120	+	90 × 2 = 300

*Step 3: Find the allele frequencies for **A** and **a**: divide the total alleles by 500

Allele frequency for **A** = 200/500 = **0.4**
Allele frequency for **a** = 300/500 = **0.6**

*Step 4: Plug the allele frequencies into the Hardy-Weinberg formula: A = p; a = q

p^2	+	$2pq$	$+ q^2$		
$(0.4)^2$	+	$2(0.4)(0.6)$	$(0.6)^2$		
0.16		**0.48**	**0.36**	**genotype frequencies (*expected*)**	

	observed	expected
AA	0.16	0.16
Aa	0.48	0.48
aa	0.36	0.36

In this example, the observed and the expected genotype frequencies are the same. Consequently, no evolution has occurred in this hypothetical population. It is important to note that mutation, natural selection, gene flow, and genetic drift regularly occur in populations because these evolutionary forces cause change in human (and nonhuman) populations. Also, human populations do not mate randomly. We choose our mates based on a myriad of biological and cultural characteristics.

Summary of Key Ideas

1. Before Mendel's work on heredity was rediscovered, blending inheritance was widely accepted as the model of inheritance where the mother's and father's hereditary units (genes) mix or "blend" to form characteristics in the offspring.

2. Mendel's experimental design was successful because he chose a pea plant (*Pisum sativum*) that had a short generation time, had male and female parts, was easy to grow, and had easily distinguishable traits. In addition, Mendel kept detailed notes on his results.
3. Dominant traits always mask recessive traits.
4. Recessive traits are expressed when the offspring inherits the recessive genes from both parents.
5. The Mendelian ratio (in the F^2 generation) is 3:1 (three purple: one white).
6. The genotypic ratio is 1:2:1 (25% homozygous dominant: 50% heterozygote dominant: 25% homozygous recessive).
7. The law of segregation states that there are two alleles in every trait: one inherited from the mother and one from the father.
8. The law of independent assortment states that the inheritance of one trait does not affect the inheritance of another trait.
9. Humans have 46 chromosomes (23 pairs) in our somatic (body) cells and 23 chromosomes in our sex cells (egg and sperm).
10. Mitochondrial DNA is different from cell to cell (heteroplasmic) while nuclear DNA is identical in all body cells (homoplasmic).
11. Regulatory genes turn other genes "on" and "off."
12. Mendelian traits are discrete—absent or present, controlled by a single locus, and, therefore, not affected by the environment. Polygenic and pleiotrophic traits are controlled by several loci, measured on a continuum, and affected by the environment.
13. The Hardy-Weinberg equilibrium states that genotype frequencies will remain constant (no evolution) if certain assumptions are met: random mating, no mutation, no natural selection, no gene flow (no immigration), and no genetic drift (population is very large).

Key Terms

Antibodies
Antigens
Autosomal chromosome
Blending inheritance
Cell
Chromosomes
Cross-fertilization
Crossing over
DNA
Diploid
Dominant
Double helix
Gregor Mendel
Gemmules
Gene
Genotype
Hardy-Weinberg equilibrium
Heterozygote
Haploid
Homozygote
Hybrids
Lactose intolerance
Law of independent assortment
Law of segregation
Locus (loci *pl.*)
Meiosis
Mendelian inheritance
Mendelian ratio
Microevolution
Mitosis
Modern synthesis
Natural selection
Phenotype
Pleiotrophic
Polygenic
Polymorphism
Punnett square
Recessive
Recombination
Regulatory genes
Transcription
Translation
True-breeding

Further Readings

Jones, S. 1993. *The language of genes: Solving the mysteries of our genetic past, present, and future.* New York: Doubleday.

Portugal, F. H., and J. S. Cohen. 1977. *A century of DNA: A history of the discovery of the structure and function of the genetic substance.* Cambridge, MA: MIT Press.

Sykes, B. 2001. *The seven daughters of Eve: The science that reveals our genetic ancestry.* New York: Norton.

Watson, J. D. 1968. *The double helix: A personal account of the discovery of the structure of DNA.* New York: Atheneum.

INTERNET RESOURCES

Mendelweb
http://www.mendelweb.org
This Web site provides information about the life and work of Gregor Mendel and the foundation of genetics.

Chapter 3: Genetics and Heredity

Study guide: By knowing the material below you will be well prepared for the exam. Don't forget to read the chapter and study your notes (this includes the chapter worksheets). You will need a calculator to answer some of the questions.

1. Be able to define Mendel's law of segregation?
2. What are chromosomes? XX (male or female?) XY (male or female?)
3. How many chromosomes are found in somatic cells and sex cells (gametes)?
4. Be able to give an example of a somatic cell
5. Be able to discuss mitochrondrial DNA
6. Define the Modern Synthesis
7. Be able to distinguish between genotype and phenotype
8. Why did Mendel choose the pea plant?
9. Define Deoxyribonucleic acid (DNA)
10. Be able to define gametes?
11. Be able to define Mendel's law of independent assortment?
12. If you have two parents with the genotype *Aa*, what is the chance that they will have an offspring with the genotype *AA*? *Know how to use a punnett square*
13. Be able to define meiosis
14. Be able to define mitosis
15. Recombination and crossing-over are very important in the genetic process because they maintain______.
16. Be able to explain crossing over in meiosis?
17. Be able to define haplotype
18. If the mRNA codon is UGG, the corresponding tRNA anticodon will be____________.
19. What is transcription and translation (as it relates to DNA)?
20. All parts of the DNA code for proteins. True/False?
21. The field of genetics concerned with the law of inheritance is known as__________?
22. What is the shape of the DNA molecule?
23. What is a codon? What does it specify (code for)?

24. Adenine-Gunanine; Thymine-Cytosine. True/False?
25. What is the cause for the anatomical differences between chimpanzee and humans despite the 98% genetic similarity or lactose intolerance in some populations?
26. Be able to define dominance and codominance
27. Be able to define polygenic and peiotrophic
28. Proteins consist of____________________.
29. Be able to define population genetics
30. Be able to explain the Hardy-Weinberg Equilibrium and list the assumptions necessary to satisfy this model.
31. Be able to list and explain the forces of evolutionary change
32. Be able to define locus
33. Be able to define point mutation
34. Be able to calculate the genotype frequencies
35. Be able to calculate the allele frequencies
36. Be able to calculate the number of alleles
37. Be able to calculate the total number alleles in a population
38. Be able to calculate the *observed* genotype frequencies
39. Hardy-Weinberg equation ($p^2 + 2pq + q^2$). Be able to calculate the *expected* genotype frequencies
40. Mutations occur at a very low rate. True/False?

NAME ________________________________ DATE ________________

Chapter 3: Genetics and Heredity

HOME WORK EXERCISES (10PTS.)

The goal of the home work exercises is assess your knowledge of the concepts and ideas presented in the chapter. Please answer the questions in each section as you go. When class lecture on this chapter is completed, carefully tear the home work exercises from the textbook and turn it in. **[Bonus pts. = 10]**

Use a pen or pencil to answer the questions below. Make sure your writing is LEGIBLE.

Section 3.1: Genetics and Heredity

1. Who is Gregor Mendel? Why is he significant?

2. Explain how Mendel's work refuted the old concept of blending inheritance.

3. Describe Mendel's experiment. Using one trait of the pea plant, flower color, what did he get in the first generation of the cross? What did he get in the second generation? What two laws were formulated as a result of Mendel's work?

4. Definitions: Put the correct answers (indicated by the letters) in the spaces on the right.

A. Skin color
B. 23 pr. Chromosomes
C. Heterozygous dominant
D. Dominant
E. True breeding
F. Homozygous dominant
G. Discrete traits
H. Homozygous recessive

_______ *As*
_______ *ss*
_______ genotype
_______ phenotype
_______ *AA*
_______ absence or presence
_______ always expressed
_______ pure

5. Using a Punnett square, solve the following problem. If a woman with genotype *AB* (blood type) mates with a man with genotype *AO* (blood type), what is the expected proportion of genotype *AB* (blood type) in the offspring.

6. Explain the modern synthesis. When was it developed?

Section 3.2: Molecular and Population Genetics

1. Define Nuclear Deoxyribonucleic Acid (nDNA). Name the scientists who discovered the structure of nDNA?

2. What are chromosomes? How many chromosomes are in the human body (somatic) cells? How many chromosomes are in the human sex cells (egg and sperm)?

3. Define mitosis and meiosis. How does meiosis relate to Mendel's principle of Segregation and Law of Independent Assortment?

4. What is the evolutionary significance of meiosis?

5. Describe the steps involved in protein synthesis.

6. Define antigens and antibodies. Use these terms to explain the ABO blood system (in other words, what does 'A' signifies for example and why do we have to receive the correct blood type?)

7. The ABO blood system may be related to disease resistance and natural selection. Discuss this hypothesis.

8. Define polygenic and pleiotropic traits. Compare these complex traits to simple Mendelian traits.

9. Discuss the Hardy-Weinberg equilibrium.

10. Solve the following the Hardy-Weinberg problems:
    A. Imagine a locus with two alleles, ***A*** and ***a***. If there are people with the following genotypes: ***AA*** **= 81;** ***Aa*** **= 18;** ***aa*** **= 1**, what is the frequency of the *A* allele?

B. Imagine a locus with two alleles, ***F*** and ***G***. If the frequency of the ***F*** allele is **0.6**, then the frequency of the ***G*** allele must be______.

C. Assume a population with the following genotypes: ***AA*** **= 25;** ***Aa*** **= 50;** ***aa*** **= 25**. How many ***A*** alleles are in the population?

25 x2 = 50 + 50

A = 100

D. Assume a population with the following genotypes: ***AA*** **= 31;** ***Aa*** **= 54;** ***aa*** **= 22**. How many total alleles in the population?

E. Assume two alleles, ***A*** and ***a*** at a given locus. If the frequency of the ***A*** **allele is 0.2** and the frequency of the ***a*** **allele is 0.**8, using the Hardy-Weinberg formula ($p^2 + 2pq + q^2$), calculate the expected genotype frequencies.

CHAPTER 4

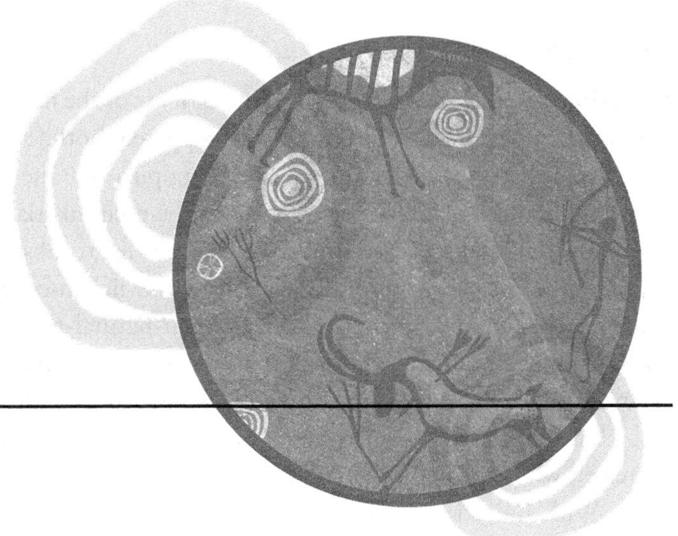

Evolutionary Forces

The evolutionary forces are mutation, natural selection, gene flow, and genetic drift. These forces interact with one another over time to increase or decrease variation. The result is evolutionary change within and between biological populations. While these forces never stop, their intensity is modified by changing cultural adaptations and population size. Now let us discuss each evolutionary force in detail and examples will be used to increase clarification.

Mutation

A **mutation** is a random change in the DNA. It is the only source of brand-new genetic variation in a population. Most mutations are deleterious; consequently, they occur at very low frequencies (0.0001) simply because most individuals with mutations die. When you die, your genes die with you. Sometimes mutations are found in high frequencies (greater than 1 percent [.01]) in populations because natural selection is keeping the mutant gene in the population. In other words, the mutation is promoting survival and reproduction. In short, mutations can be deleterious, advantageous, and neutral. Let us look at two examples of genetic mutations. First, **point mutation** is a random change of one nucleotide base to another. This change might be deleterious, advantageous, or neutral (synonymous). **Nucleotide bases** work in groups of three (triplet) or **codons** (Figure 4.1). These codons code from specific amino acids. **Amino acids** are subunits of **proteins**. During **protein synthesis**, amino acids are linked in a chain to form proteins. Consequently, a mutation is deleterious because it changes the amino acid and affects the protein. A good example is sickle cell disease. The random point mutation changes **glutamic acid** (the correct amino acid) to valine (another amino acid), which results in a sickle-shaped RBC. The sickle cell gene is discussed in detail later in the chapter.

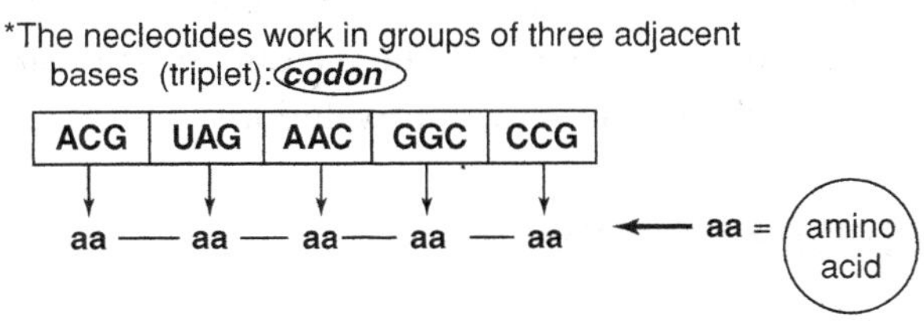

Figure 4.1. A codon is a nucleotide triplet that codes for specific amino acids.

Different codons code for the same amino acid. This redundancy explains the neutral aspect of mutations. A mutation is advantageous when it promotes survival and reproduction in a population. Second, **frameshift mutation** is a random change—addition or deletion—of one or more nucleotide bases resulting in a flawed mRNA that is translated incorrectly during protein synthesis.

Natural Selection

In Chapter 2, natural selection was defined as survival and reproduction for those individuals with adaptive traits *in a given environment.* There are, however, different types of selection in nature. In this text, we will discuss (1) directional selection; (2) stabilizing selection; and (3) balancing selection.

DIRECTIONAL SELECTION

Directional selection is selection for one extreme or the next in a continuous trait. A classic example is the peppered moth (Figure 4.2). There are two varieties of moths in this population: the peppered moth and the dark-colored moth. In the preindustrial environment, the moths rested on lichen-covered tree trunks. The peppered moths were camouflaged against the lichen-covered trunks and, therefore, not easily seen by their natural predator, birds. In contrast, the dark-colored moths were very conspicuous against the lichen-covered tree trunks and were subsequently eaten by birds. Over time, the frequency of the dark-colored variety decreased while the peppered-colored variety thrived.

With industrialization came pollution, which eventually killed the lichen. Soon, postindustrial tree trunks were covered with soot. In this environment, the peppered moth was now conspicuous against the background of soot-covered tree trunks, while the dark-colored moth was camouflaged. The peppered moths were eaten while the dark-colored moth population rebounded. An important point to take from this example is that *variation* is critical to survival in changing environments. The moth population would have become extinct if there were only one variety. Another example is the increase in brain size over 4 million years in human evolution.

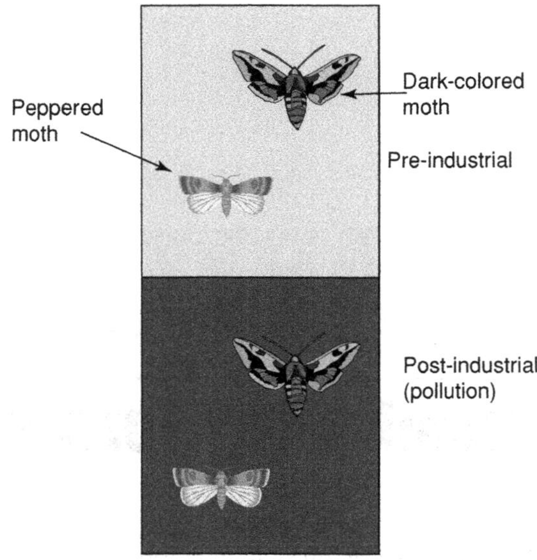

Figure 4.2. Natural selection in moth species in pre- and postindustrial environments: Peppered moth variety has a higher frequency in preindustrial environments and the dark-colored variety has a higher frequency in postindustrial environments.

STABILIZING SELECTION

Stabilizing selection is selection against the extremes (higher or lower) in a continuous trait. An interesting example is human birth weight (Figure 4.3). The optimal birth weight of babies born in first (England) and second world (United States) countries, for example, is 5 to 7 pounds. The mortality rate increases for babies born too small or too large. In addition, childbirth becomes very difficult for mothers whose babies are too big. Consequently, there is evolutionary selection against the extremes for the average (in a given environment). Of course, 5 to 7 pounds will not be the optimal birth weight in populations where there is starvation and disease or in populations living at high altitude in the Andes (Peru) and the Himalayas (Tibet). Remember, traits are relative; what works in one environment may not work in another.

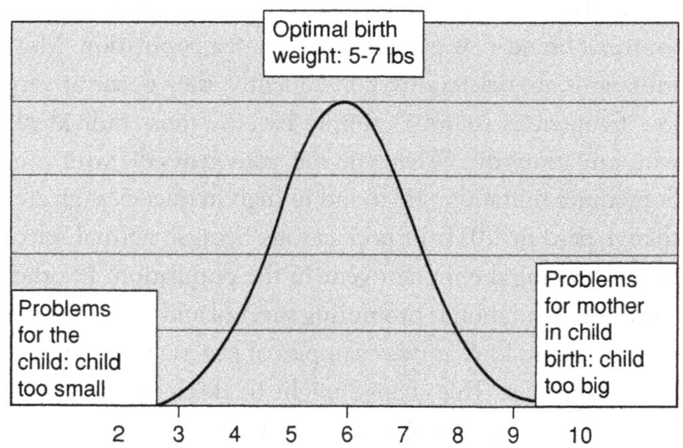

Figure 4.3. Stabilizing selection: selection against the extremes in a continuous trait.

BALANCING SELECTION

Balancing selection is selection against the homozygotes and for the heterozygotes in a discrete trait. A classic example is the **malaria–sickle cell hypothesis.** Sickle cell is a random point mutation in the beta chain of the hemoglobin molecule resulting in sickle-shaped RBCs. It is very different from the normal RBC. Because of its sickle shape, it restricts normal blood flow by clogging the arteries, preventing oxygenated blood from supplying vital regions of the body—hence the term **sickle cell anemia** (hemoglobin^{s}-hemoglobin^{s}). The prognosis is not good for children with sickle cell anemia (Mendelian inheritance: sickle cell allele inherited from both parents). Fortunately, individuals with the trait (heterozygotes) survive to adulthood because they only have a low rate of RBC sickling.

Thousands of years of slash-and-burn agriculture in the tropical areas of sub-Saharan Africa, the Mediterranean, Asia, and Indonesia has created optimal environments for female mosquitoes—particularly, the anopheles mosquito—to lay their eggs. Slash-and-burn agriculture is characterized by felling and burning trees to clear land for temporary agriculture. In a short time period, the land becomes useless and the farmer moves on to a new area. During the rainy season, the inactive land becomes like a sponge, soaking up and holding standing water. Subsequently, anopheles mosquitoes carrying the malaria parasite lay their eggs. Soon, the mosquito population increases, and then they transmit the malaria parasite to the local human populations (Figure 4.4).

It is interesting that where you find malaria you find the **sickle cell gene** in high frequency. Why? What is the connection between the sickle cell gene and malaria? The malaria parasite cannot survive in a sickle-shaped RBC; it needs a normal RBC to complete its cycle of life. This is advantageous for individuals with the sickle cell trait because they do not suffer the deleterious effects of malaria. In short, these heterozygotes survive and reproduce, passing the sickle cell gene from one generation to the next so that it increases in frequency. In essence, it is these heterozygotes who perpetuate the sickle cell gene simply because it promotes survival in a malarial environment. They have the highest reproductive fitness. Normal individuals die of malaria and individuals with both copies of the sickle cell gene (Hb^SHb^S) die of sickle cell anemia (Figure 4.5). In genetics, this is called a **balanced polymorphism**.

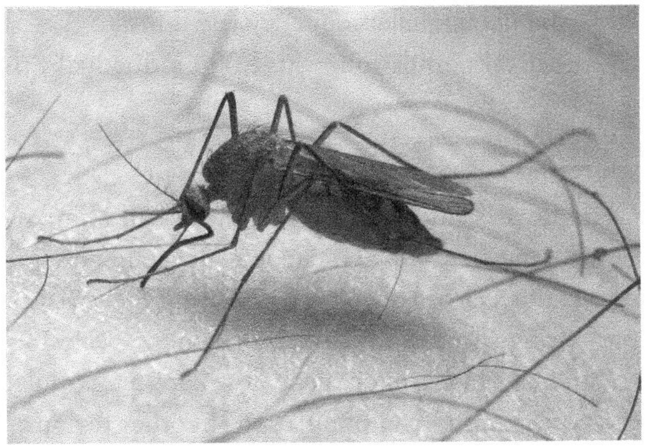

Figure 4.4. An anopheles mosquito in the process of biting a human being.

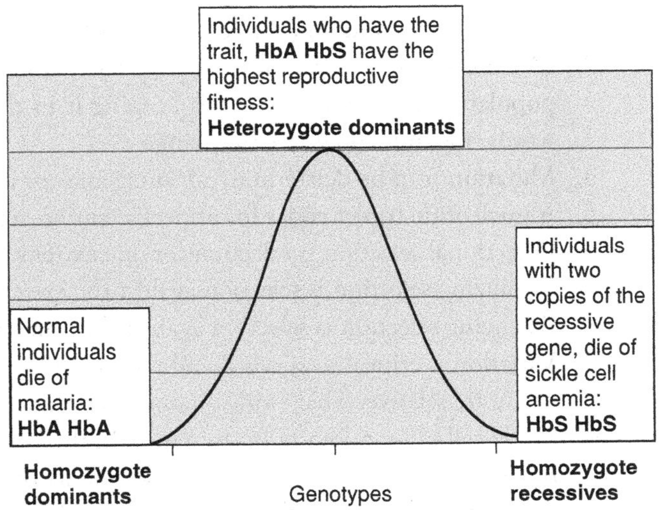

Figure 4.5. Balancing selection: selection against the homozygotes for the heterozygotes.

Genetic Drift

Genetic drift is the most complex of the evolutionary forces. It is a random change (chance event) in allele frequency from one generation to the next. It has its greatest effect in small populations, and it removes variation leading to **homogeneity** given enough time. Examples of genetic drift can be found in isolated or island human and nonhuman populations simply because there is a limited number of mates in isolated areas and no input of genes from the outside (immigration) to add variation. Therefore, you have increased homogeneity and, to some extent, inbreeding. Here is a simple example. One hundred Bloomsburg University students (50 with brown eyes and 50 with blue eyes) are on a small ship in the Pacific Ocean. Let us use 'A' allele = brown eye (0.5) and 'a' allele = blue eye (0.5). Suddenly, a tsunami appeared causing the ship to capsize. The brown-eyed students were able to swim to a deserted island. The blue-eyed students all drowned. First, the 50 brown-eyed students would be the **founder** population on that deserted island. And, from generation to generation there will be brown-eyed individuals (homogeneity). Looking at the alleles, the initial frequencies were A = 0.5 and a = 0.5 for a total = 1.0 (100 percent). After this chance event or accident, there is only the A allele or 100 percent A. This is a very large change in allele frequency (A = 0.5 to A = 1.0) because the population was small (100 individuals). If this hypothetical population were 1,000 individuals, then the allele frequency change would not be so drastic.

Gene Flow

Gene flow is migration (immigration) or admixture. Specifically, it is the movement of genes from one population to another. Unlike genetic drift, gene flow increases the variation within populations bringing the respective populations closer together (decreasing the variation between them). In human evolution, gene flow was critical in preventing speciation (formation of new species) in the widely dispersed *Homo sapiens* in Africa, Europe, Asia, and Indonesia. There is more gene flow between populations that live in same regions; therefore, they generally resemble each other more than populations in other regions, that is, Europeans, Asians, sub-Saharan Africans, to name a few.

Summary of Key Ideas

1. The evolutionary forces mutation, natural selection, genetic drift, and gene flow interact with one another over time to increase or decrease variation within and between biological populations.
2. Mutation is a random change in the DNA. It is the only source of brand-new variation.
3. Mutations occur in very low frequencies (0.0001). When found in high frequencies (greater than.01) in populations, natural selection is keeping it in the population because it is promoting survival; in other words, the individuals are surviving.
4. Mutations can be deleterious, advantageous, or neutral.
5. A nucleotide triplet codes for a specific amino acid.
6. Directional selection is selection for one extreme or the next in a continuous trait.
7. Stabilizing selection is selection against the extremes (higher or lower) in a continuous trait.
8. Balancing selection is selection against the homozygotes and for the heterozygotes in a discrete trait.
9. Variation is critical to survival and reproduction in changing environments.
10. Traits are relative: what works in one environment may not work in another environment.
11. Sickle cell is a random point mutation in the hemoglobin molecule.
12. An individual with sickle cell anemia inherits the sickle cell gene from both parents. In contrast, an individual with the sickle cell trait inherits the sickle cell gene from only one parent.
13. Malaria is a parasite carried by the anopheles mosquito (the vector).
14. Fitness in an evolutionary sense refers to the probability of survival and reproduction.
15. Genetic drift is a random change in allele frequency from generation to generation. It is a change event that changes allele frequencies removing variation. Its greatest effect is on small populations.
16. Gene flow is admixture or migration. Genes are exchanged between populations, which increases the variation within the respective populations, decreasing the variation between them.

Key Terms

Amino acid
Balanced polymorphism
Balancing selection
Codon
Directional selection
Founder effect
Frameshift mutation
Gene flow
Genetic drift
Glutamic acid
Homogeneity
Malaria–sickle cell hypothesis
Mutation
Nucleotide bases
Point mutation
Proteins
Sickle cell anemia
Sickle cell gene
Stabilizing selection
Synonymous mutation

Further Readings

Kettlewell, H. B. 1973. *The evolution of melanism.* Oxford, UK: Oxford University Press.

Livingstone, F. B. 1958. Anthropological implications of sickle cell gene distribution in West Africa. *American Anthropologist* 60: 533–62.

Molnar, S. 2002. *Human variation,* 5th ed. Upper Saddle River, NJ: Prentice Hall.

Ridley, M. 2004. *Evolution.* Malden, MA: Blackwell Science.

INTERNET RESOURCES

Natural Selection
http://en.wikipedia.org/wiki/Natural_selection
This page in Wikipedia describes the process of natural selection and other evolutionary forces.

The Talk Origins Archive
http://www.talkorigins.org
This Web site provides information on genetics, evolutionary processes, extinction, and many other topics related to human evolution.

Chapter 4: Evolutionary Forces

Study guide: By knowing the material below you will be well prepared for the exam. Don't forget to read the chapter and study your notes (this includes the chapter worksheets).

1. Be able to list and explain the forces of evolutionary change
2. Be able to define point mutation and synonymous point mutation
3. Be able to define and give an example of each: directional selection; stabilizing selection; and balancing selection
4. Based on the text, what is the possible cause for the significant increase in the incidence of malaria over the centuries?
5. Other names for gene flow are________________________________
6. Gene flow___________________variation *within* populations and___________________variation *between* populations.
7. Be able to define microevolution and macroevolution
8. Be able to define genetic drift and Founder effect
9. Genetic drift___________________variation *within* populations and___________________variation *between* populations.
10. Be able to define *balanced polymorphism*

NAME ______________________________ DATE ______________

Chapter 4: Evolutionary Forces

HOME WORK EXERCISES (10PTS.)

The goal of the home work exercises is assess your knowledge of the concepts and ideas presented in the chapter. Please answer the questions in each section as you go. When class lecture on this chapter is completed, carefully tear the home work exercises from the textbook and turn it in. **[Bonus pts. = 5]**

Use a pen or pencil to answer the questions below. Make sure your writing is LEGIBLE.

Section 4.1: Mutation

1. List and define the two types of mutation mentioned in the readings. Explain the following statement: mutations can be deleterious, advantageous, and neutral.

2. Mutations are the only source of______________________________.

3. Mutations____________variation within populations and____________variation between populations.

Section 4.2: Natural Selection

1. Using your own example, explain how natural selection works.

2. Explain *directional, stabilizing,* and *balancing* selection. Stabilizing and balancing selection are similar, but what is the key difference between them.

3. Why do the individuals with the sickle cell trait (heterozygotes) have the highest "fitness" in a malarial environment?

4. Natural selection______________________________variation within populations and______ _______________________variation between populations.

Section 4.3: Genetic Drift

1. Define genetic drift. Explain how a founder population might be formed.

2. Genetic drift is an evolutionary force that is most powerful when acting on____________________ populations.

3. Genetic drift____________variation within populations and_______________variation between populations.

Section 4.4: Gene flow

1. Define gene flow. Explain why gene flow is integral to the concept of species.

2. Gene flow____________variation within populations and_______________variation between populations.

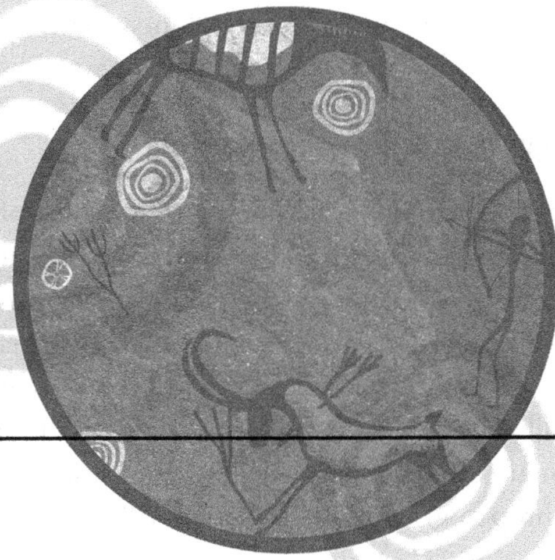

CHAPTER 5

Primate Evolution

Primate evolution is very complex because it goes back 65 million years, and at different periods during this long history, there have been 7 to 8 million year gaps in the primate fossil record. In addition, a large percentage of the thousands of ancient primate fossils discovered in Africa, Europe, Asia, and North and South America are small fragments of bone and teeth. Despite this complexity, we study primate evolution because it gives us some insight into the origins and evolution of prosimians and major anthropoid groups, such as monkeys, apes, and humans (see Chapter 6) [Figure 5.1].

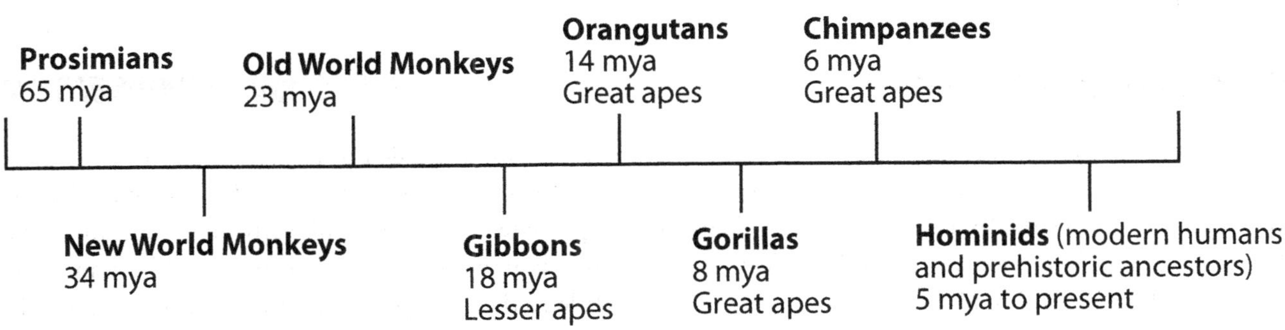

Figure 5.1. Primate Evolutionary Tree showing different divergence times for major primate groups.

The Hypotheses for Primate Origins

There are three major hypotheses for primate evolution. The traditional hypothesis or **arboreal hypothesis** (or life in the trees) was formulated over a century ago by British anatomists Sir Grafton Elliot Smith (1871-1937) and Frederic Wood Jones (1879-1954). It has been, and, to some extent, still is the most influential hypothesis in the field of primatology. However, in the last 40 years the arboreal hypothesis has been challenged by two creative and interesting hypotheses: the **visual predation hypothesis** and the **angiosperm radiation hypothesis**.

ARBOREAL HYPOTHESIS

In this hypothesis, Smith and Jones hypothesized that forward-facing eyes, stereoscopic-binocular vision providing distance and depth perception, color vision, and grasping hands and feet, all evolved for adaptation to life in the trees (see Chapter 6). On the surface, this hypothesis seems perfectly valid. But upon deeper examination, it is a weak hypothesis for primate origins because many mammals such as cats, opossums, sloths, treeshrews, and squirrels, to name a few, can climb trees if not totally arboreal. This weakness led to the proposition of a stronger hypothesis by anthropologist Matt Cartmill (1974).

VISUAL PREDATION HYPOTHESIS

While Cartmill concedes that components of the arboreal hypothesis is generally true, in this hypothesis he theorized that forward-facing eyes, stereoscopic-binocular vision providing distance and depth perception, color vision, and grasping hands and feet all evolved for preying on insects and small animals in an arboreal environment. Specifically, forward-facing eyes, like the big cats on the African savanna, enhance focus on the prey; grasping hands and feet secure the animal as it moves in to grab the prey. The problem here is that not all primates eat insects and small animals. But all primates eat fruits. In short, Cartmill's hypothesis does not consider the relationship between primate characteristics and the acquisition of fruit. This gap in knowledge was filled by a hypothesis proposed by anthropologist Robert Sussman (1991).

ANGIOSPERM RADIATION HYPOTHESIS

In this hypothesis, Sussman hypothesized that forward-facing eyes, stereoscopic-binocular vision providing distance and depth perception, color vision, and grasping hands and feet all evolved for eating fruits made available with the radiation of flowering plants called angiosperms. Specifically, forward-facing eyes and color vision enabled the animal to not only find fruits within the thick canopy but also to note, by coloration, if they were ripe. And, grasping hands and feet enabled the animal to hold on while it picked the fruits.

We will never know which hypothesis accurately explains why primates evolved. But together, all three hypotheses provide important components of what actually happened 65 million years ago. Now, let us look at the fossil evidence for primate evolution.

Proprimates: First Primates?

By 65 million years ago, the dinosaurs were extinct (see Chapter 7 for a detailed discussion on earth's geological history). This allowed for the proliferation and diversification of mammals ushering in the **Cenozoic** era/**Paleocene** epoch or simply the "Age of Mammals." Whether or not primates were part of this early diversification in the Paleocene (65 to 54 million years ago) is unclear. Traditionally, a highly diverse group of "primate-like mammals" technically known as **plesiadapiforms**, were given the lofty title as the "first primates." But with the advances in techniques and scientific analysis within the field, plesiadapiforms were not good candidates for the first primates because they did not have forward-facing eyes and grasping hands and feet. In addition, they did not have other more complex primate traits such a **postorbital bar**, **postorbital closure**, **auditory bulb**, or large brain-body size ratio. Rather than being the first primates, they might be a lineage leading to the primate order or **proprimates** because one plesiadapiform species, ***Carpolestes simpsoni***, whose skeleton was discovered in Wyoming's Bighorn Basin, had general primate traits such as grasping hands and feet, and nails instead of claws (Gingerich 1986,1990; Larsen 2008). The *Carpolestes* specimen has major implications for hypotheses of euprimate origins.

Euprimates: The First True Primates

At approximately 56 million years ago, technically known as the **Eocene** epoch (56 to 34 million years ago), there was global warming (see Chapter 7). Most scholars believe that the "first true primates" or euprimates—a term coined by the prominent evolutionary primatologist Elwyn Simons (b. 1930) to mean primates of modern aspect—appeared at this time. They had the general primate traits such as forward-facing eyes, grasping hands and feet, postorbital bar, nails instead of claws, and large brain-body size ratio. Euprimates are generally placed in two closely related groups: the **adapids** and the **omomyids** found in the western United States, western Europe, Africa, and Asia (Nystrom and Ashmore 2008; Larsen 2008). In terms of size, adapids were similar to modern lemurs and omomyids to tarsiers (see Chapter 6).

As the Eocene epoch came to a close, there was global cooling which might have led to the extinction of many plant and animal species including many of the euprimates (Nystrom and Ashmore 2008; Larsen 2008). Related to this are two key questions. First, which groups of euprimates were ancestral to **anthropoids** (monkeys, apes, and humans)? The adapids and omomyids are likely candidates according to some researchers (Gingerich 1976; Szalay et al. 1987). Second, is Africa or Asia the ancestral home of anthropoids? Two fossils, ***Eosimias***, found in China and dated to about 42 million years ago, and ***Biretia***, found in North Africa and dated to 37 million years ago, have skeletal traits that are similar to later fossil and living anthropoids. For instance, *Eosimias* has a calcaneus bone (or heel bone) that is similar in shape and structure as in anthropoids (Larsen 2008). And *Biretia* has premolars (bicuspids) in its lower jaw that are structured like anthropoids. The earlier time period of *Eosimias* would suggest that anthropoids originated in Asia. However, a larger sample of fossils dating to 42 million years must be found in Asia in order to accept an Asian ancestry for anthropoids. The schematic in figure 5.2 gives a visual depiction of the discussion in this paragraph.

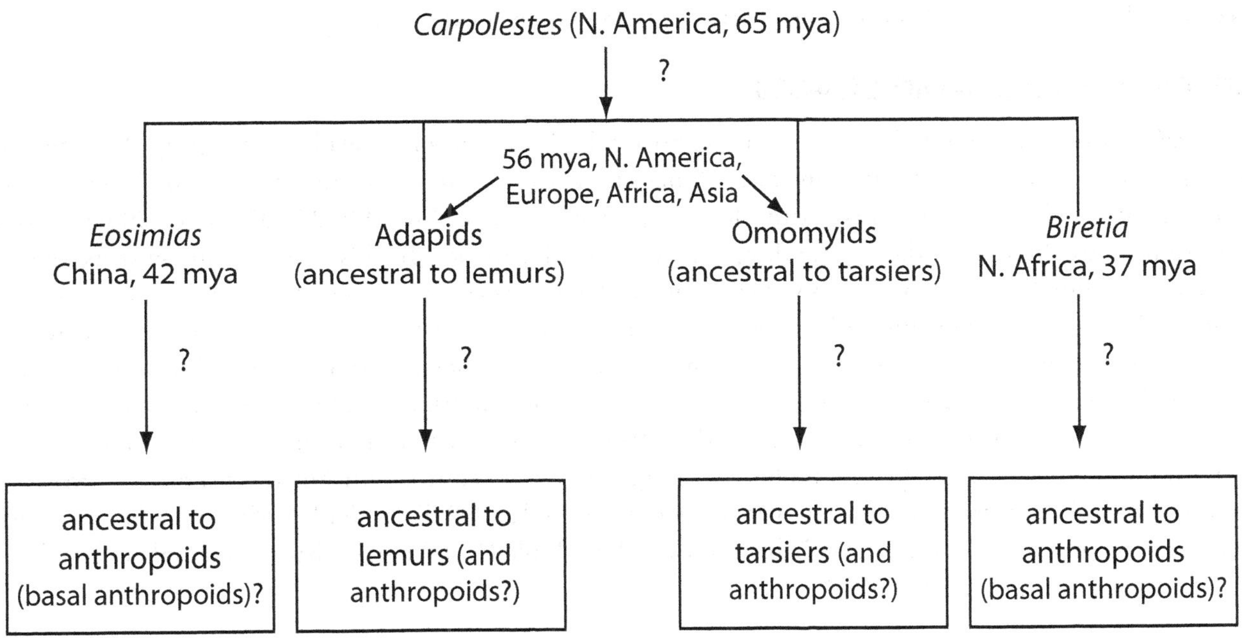

Figure 5.2. Schematic of the first true primates and the origins of anthropoids.

Anthropoids: The Higher Primates

Global cooling in the **Oligocene** epoch (34 to 24 million years ago) led to the evolution of three groups of higher primates in North Africa: **oligopithecids**, **parapithecids**, and **propliopithecids**. Parapithecids are believed to be ancestral to **platyrrhines** (New World monkeys) because they have three premolars on each side of the mouth—upper and lower—for a total of 12 and New World monkeys have the same (see Chapter 6) [Nystrom and Ashmore 2008; Larsen 2008]. In contrast, propliopithecids are believed to be ancestral to **catarrhines** (Old World monkeys, apes, and humans) because they have two premolars on each side of the mouth—upper and lower—for a total of 8 and Old World monkeys have the same. Figure 5.3 shows the divergence of Oligocene anthropoids into platyrrhines and catarrhines. Now, let us delve further into the origins of New World monkeys.

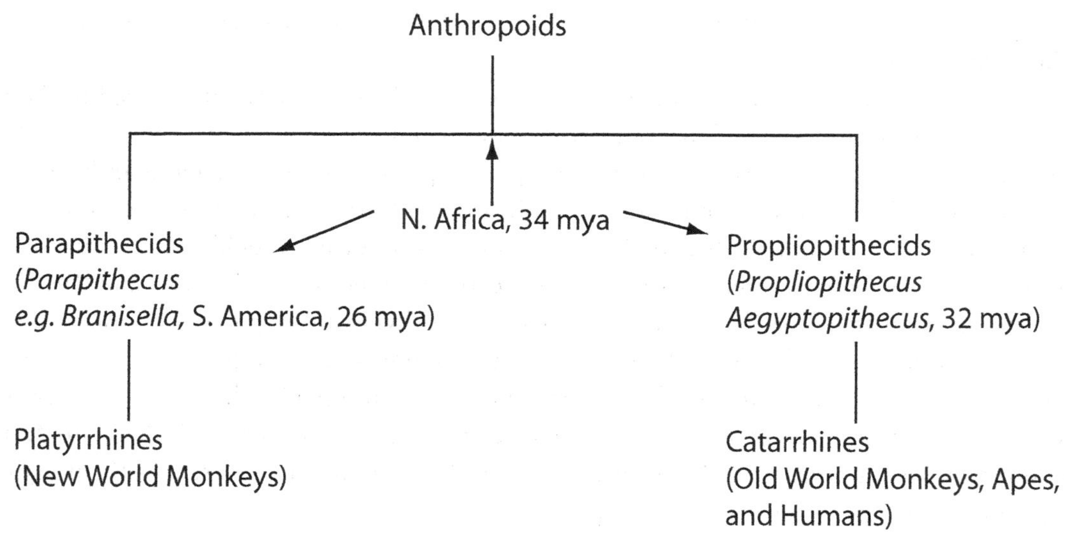

Figure 5.3. The divergence of Oligocene anthropoids into platyrrhines and catarrhines.

PLATYRRHINE ORIGINS: NEW WORLD MONKEYS

Branisella boliviana, from the late Oligocene deposits in Bolivia, is the oldest evidence of a platyrrhine monkey. *Branisella* shares ancestral traits with living platyrrhines such as three premolars in each quadrant of the mouth for a total of 12 and molars in the upper jaw with a four-cusp chewing surface (Relethford 2005; Larsen 2008; Nystrom and Ashmore 2008). So, how did early anthropoids get to South America? By this time (34 million years ago), the African and South American continents were separated. Many years ago, it is thought that platyrrhines and catarrhines evolved by **parallel evolution** from prosimians. Remember, Eocene euprimates were found in the western United States. However, anatomical similarities and DNA evidence showing strong relationship between platyrrhines and catarrhines make it unlikely that there was parallel of platyrrhines and catarrhines. The most popular hypothesis is that the first platyrrhines originated in Africa (because there is no evidence to-date of anthropoids in North or South America before 34 million years ago) and migrated to South America via ocean currents, on "floating islands" or natural rafts of accumulated vegetation (Houle 1999; Relethford 2005; Larsen 2008; Nystrom and Ashmore 2008). Many animals probably drowned, but Houle (1999) argues that some monkeys could have survived long enough to reach South America.

CATARRHINE ORIGINS 1: OLD WORLD MONKEYS

Aegyptopithecus zeuxis is the most famous of the propliopithecids, dating between 32 and 29 million years ago (Figure 5.4). It is the largest (13 to 18 lbs) of the fossil primates from North Africa within the time range of 37 to 29 million years ago. Listed below are several key characteristics:

1. Sagittal crest
2. Large brain-to-body size (compared to other Oligocene primates)
3. Dental formula: 2-1-2-3 (one side of the mouth—upper or lower/left or right)
4. All limbs the same size

Is *Aegyptopithecus* the common ancestor of all later catarrhines? The consensus in the field is yes, even though no anthropoid fossils have been found from the late Oligocene (29 million years ago) to the early **Miocene** epoch (22 million years ago), where we see the divergence of Old World monkeys on one hand and the emergence of the **proconsulids** or Miocene apes on the other. As far as the apes are concerned, similarities in dental formula (2-1-

2-3) and incisor shape (broad and flat) between *Aegyptopithecus* and the later proconsulids indicate a possible ancestral-descendant relationship (Larsen 2008).

Global warming began in the late Oligocene and continued into most of the Miocene (23 to 5 million years ago). The Old World monkey lineage or **Cercopithecidae** (see Chapter 6) was very successful in the Miocene and would continue to proliferate into the **Pliocene** (5 to 1.8 million years ago) and **Pleistocene** (1.8 million to 10,000 years ago), eventually evolving another subfamily called **Colobinae** (see Chapters 6 and 7) due to environmental changes. The primitive cercopithecoids are called **Victoriapithecids** and the genus *Victoriapithecus* is believed to be ancestral to Old World monkeys (Benefit and McCrossin 1991; Nystrom and Ashmore 2008; Larsen 2008).

Figure 5.4. The Oligocene anthropoid *Aegyptopithecus*.
Duke University Primate Center. © 1985 David L. Brill

CATARRHINE ORIGINS 2: APES

The proconsulids, like the cercopithecoids, were very diverse in the early Miocene. After all, they lived in a lush, tropical environment. Listed below are a few key characteristics that make them candidates for the last common ancestors of all **hominoids** (apes and humans) [see Chapter 6]:

1. Large brain-to-body ratio
2. Dental formula: 2-1-2-3 (one side of the mouth—upper or lower/left or right)
3. **Skull and teeth like apes (Y-5 molar cusps pattern—between the five cusps, the grooves form a "Y" pattern; wide round cusps)**
4. All limbs the same size
5. **Limited wrist mobility**
6. **Large big toe**

Proconsul was probably an arboreal quadruped in the throes of evolutionary change (Figure 5.5).

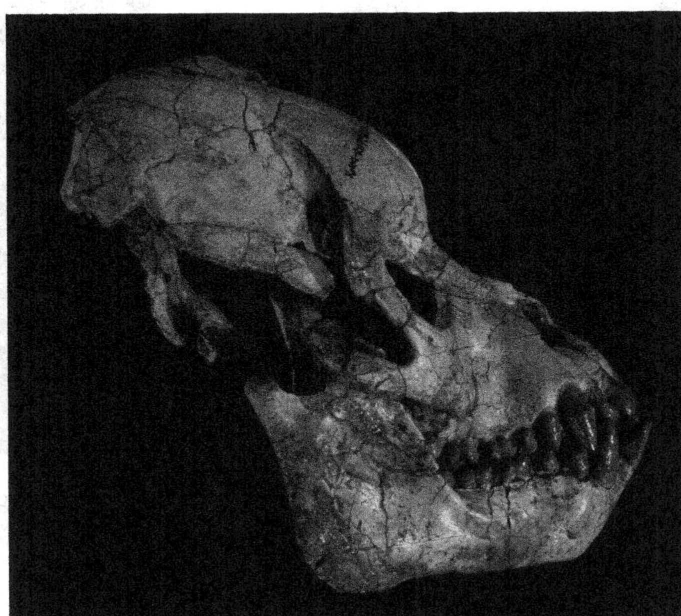

Figure 5.5. The early Miocene ape *Proconsul*.
Photo courtesy of Dr. Alan Walker.

Dryopithecines: The Mid-Miocene Apes

By the mid-Miocene (13 million years ago), a new group of apes called **Dryopithecids** had migrated into Europe and Asia. Listed below are a few key characteristics that link them to living apes:

1. Skull and brain size similar to living chimps

2. Y-5 molar cusps pattern (Y-5 molar cusps pattern—between the five cusps, the grooves form a "Y" pattern; wide round cusps)
3. Grasping feet
4. Longer arms than legs
5. Large body size (estimated from bones)

At this time, the ecology in Europe and Asia were tropical rain forests with many varieties of food based on reconstruction of ancient environments (see Chapter 7). Consequently, it is not surprising that several Miocene ape fossil specimens were found in several regions of Europe and Asia. For instance, the European Dryopithecids, ***Ouranopithecus*** and ***Oreopithecus*** were found in Greece and Italy and the well-known Asian Dryopithecid, ***Sivapithecus***, was found in Indonesia. Interestingly, the *Sivapithecus* skulls are very similar to those of living orangutans indicating a strong ancestral-descendant relationship (see Figure 5.6). This skeletal similarity is the reason why orangutans are sometimes called "living fossils."

In conclusion, primate evolution is complex. Just study the tree diagram of catarrhine evolution in Figure 5.7. Nonetheless, by studying primate evolution, we are able to formulate hypotheses about the origins of Old and New World monkeys and apes. More importantly, we are able to answer questions about the "missing link" between humans and chimps. This topic will be addressed in Chapter 8.

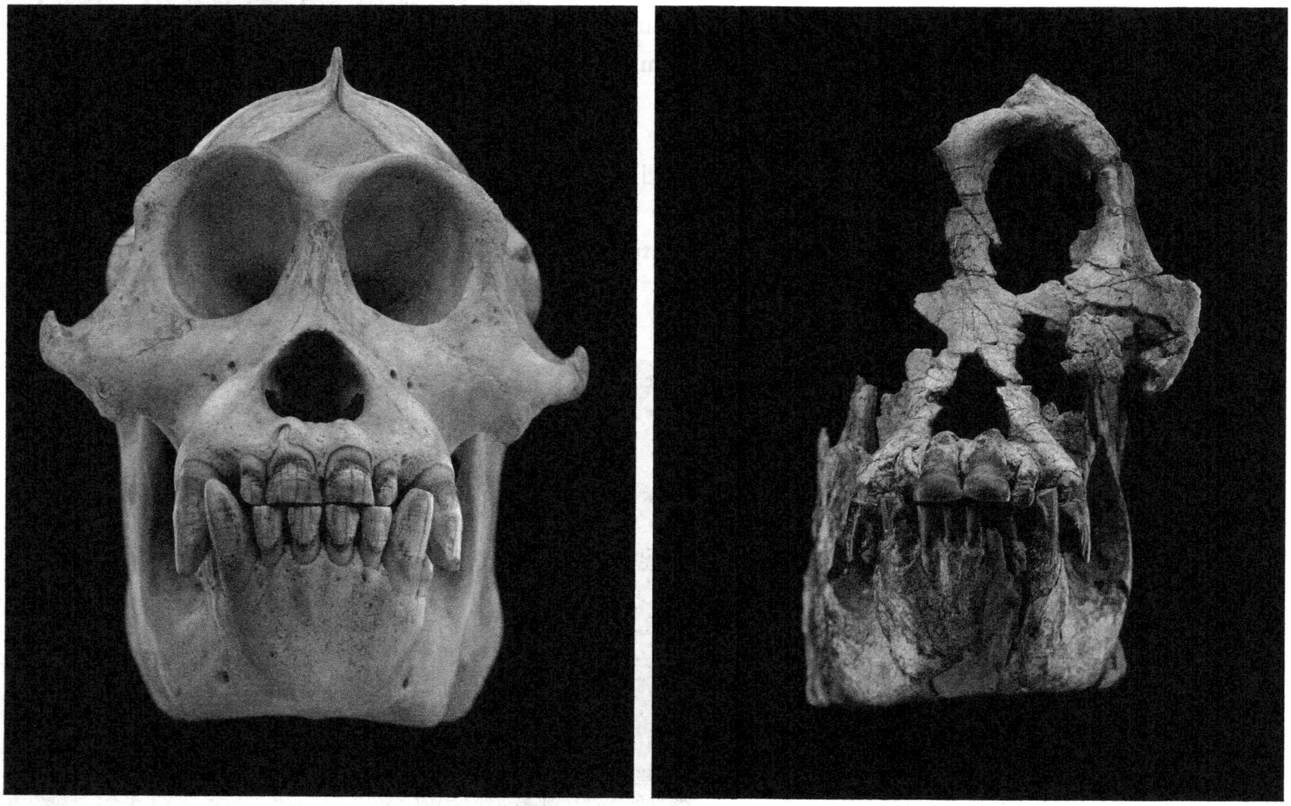

Figure 5.6. Comparison of skulls of *Sivapithecus* and a modern orangutan.
Orangutan image: Division of Mammals, National Museum of Natural History (NMNH), Smithsonian Institution. © 1996 David L. Brill
Sivapithecus image: Peabody Museum, Harvard University. © 1985 David L. Brill

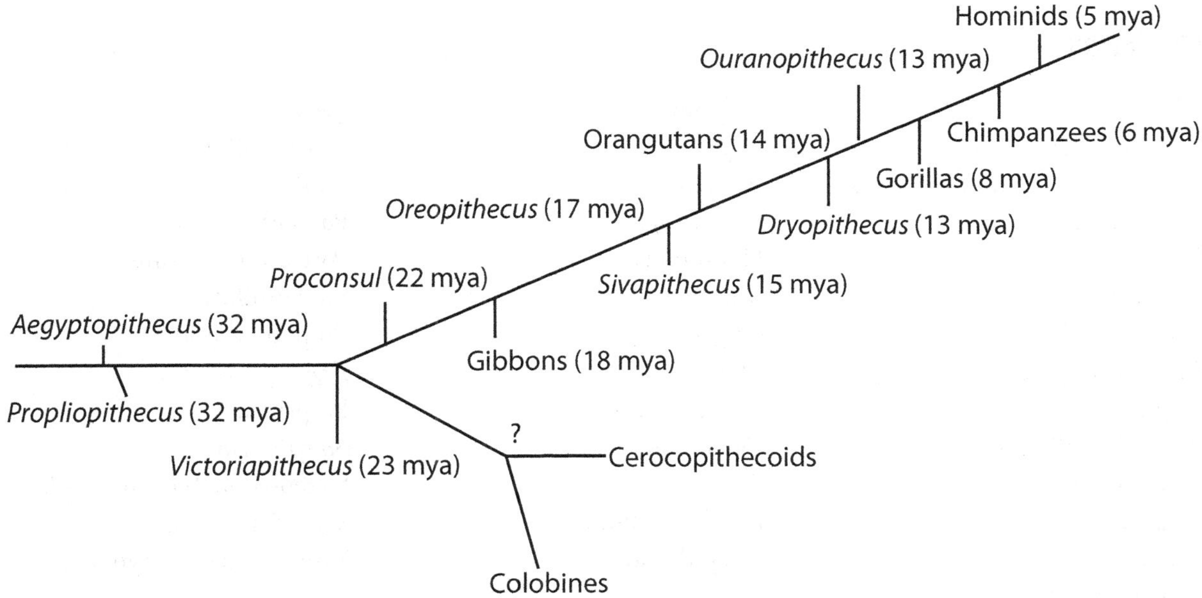

Figure 5.7. A tree diagram showing the evolution of catarrhines. Modified from Fleagle (1999) and Larsen (2008).

Summary of Key Ideas

1. Primate evolution goes back 65 million years and ancient primate fossils have been found in Africa, Europe, Asia, and North and South America.
2. Primate evolution gives us insight into the origins and evolution of prosimians, monkeys, apes, and humans.
3. There are three competing hypotheses for primate origins: the arboreal hypothesis states that forward-facing eyes, and grasping hands and feet evolved for life in the trees; the visual predation hypothesis states that forward-facing eyes, and grasping hands and feet evolved for hunting insects in an arboreal environment; and the angiosperm radiation hypothesis states that forward-facing eyes, and grasping hands and feet evolved for finding and eating fruits made available with the radiation of flowering plants.
4. Plesiadapiforms (65 mya) were "primate-like mammals" or proprimates but probably not the first primates because they did not possess the general primate traits, such as large brain-to-body size ratio, forward-facing eyes, grasping hands and feet, and nails instead of claws.
5. Euprimates (56 mya) were most likely the first primates because they possessed the general primate traits: large brain-to-body size ratio, forward-facing eyes, grasping hands and feet, and nails instead of claws.
6. The euprimates are placed in two closely related groups: the adapids and the omomyids. They are likely candidates for the ancestry of lemurs, tarsiers, and anthropoids.
7. The Oligocene (34 to 24 mya) primate groups are oligopithecids, propliopithecids, and parapithecids. The parapithecids have a dental formula of 2-1-3-3 and might be ancestral to New World monkeys. The propliopithecids, with *Aegyptopithecus* as the most well-known example, have a dental formula of 2-1-2-3. They might be ancestral to all catarrhines (Old World monkeys, apes, and humans).
8. The divergence of platyrrhines and catarrhines took place approximately 23 mya.
9. Similarities in skull and teeth between *Aegyptopithecus* and later early Miocene apes, *Proconsulids* (22 mya), indicate an ancestral-descendant relationship.
10. Fossil remains of the mid-Miocene apes, *Dryopithecus* (13 mya), were found in Europe and Asia, in addition to Africa.

Key Terms

Adapis
Aegyptopithecus zeuxis
Angiosperm radiation hypothesis
Anthropoids
Arboreal
Arboreal hypothesis
Auditory bulb
Basal anthropoids
Biretia
Branisella boliviana
Carpolestes simpsoni
Catarrhines
Cenozoic
Cercopithecidae
Colobinae
Dryopithecids
Eosimias
Euprimates
Hominoids
Miocene
Oligocene
Omomyids
Oreopithecus
Ouranopithecus
Paleocene
Parallel evolution
Parapithecids
Platyrrhines
Pleistocene
Plesiadapiforms
Pliocene
Post-orbital bar
Post-orbital closure
Proconsulids
Propliopithecids
Propilopithecus
Proprimates
Sivapithecus
Stereoscopic-binocular vision
Victoriapithecids
Visual predation hypothesis

Further Readings

Beard, C. 2004. *The Hunt for the Dawn Monkey: Unearthing the Origins of Monkeys, Apes, and Humans.* Berkeley: University of California Press.

Conroy, G. C. 1990. *Primate Evolution.* New York: Norton.

Fleagle, J. G. 1999. *Primate Adaptation and Evolution.* San Diego: Academic Press.

Miller, E. R., G. F. Gunnell, and R. D. Martin. 2005. Deep time and the search for anthropoid origins. *Yearbook of Physical Anthropology* 48: 60-95.

Rose, K. D. 1994. The earliest primates. *Evolutionary Anthropology* 3: 159-173.

Chapter 5: Primate Evolution

Study guide: By knowing the material below you will be well prepared for the exam. Don't forget to read the chapter and study your notes (this includes the chapter worksheets).

1. Be able to discuss the three hypotheses for primate origins
2. Know the skeletal characteristics of proprimates and euprimates
3. In what geographical regions of the world were proprimates and euprimates found? In what geological epoch (include dates) are they placed?
4. Euprimates are placed in two closely related groups: What are these two groups? The two groups of euprimates are most similar to modern_________________and________________________.
5. Name the first anthropoids and list the geological dates (include the epoch) and region in which they were found. What are the key traits of these fossils? Do their locations have any implication for the origins of the higher primates?
6. List the names, geological epoch, and dates of the three groups of higher primates. Including the dates of divergence, which groups might be ancestral to platyrrhines and catarrhines?

7. What is Branisella boliviana? How did anthropoids get to the New World? Discuss the various hypotheses detailed in the text.
8. What is Aegyptopithecus zeuxis? List the date, geological epoch, and four traits of Aegyptopithecus zeuxis. Why is Aegyptopithecus zeuxis important?
9. What group do the proconsulids represent? List the date, geological epoch, and four traits of the proconsulids. Why are they important?
10. What group do the dryopithecines represent? List the date, geological epoch, and four traits of the dryopithecines. Why are they important?

NAME ______________________________ DATE ______________

Chapter 5: Primate Evolution

HOME WORK EXERCISES (10PTS.)

The goal of the home work exercises is assess your knowledge of the concepts and ideas presented in the chapter. Please answer the questions in each section as you go. When class lecture on this chapter is completed, carefully tear the home work exercises from the textbook and turn it in. **[Bonus = 10pts]**

Use a pen or pencil to answer the questions below. Make sure your writing is LEGIBLE.

Section 5.1: The Hypotheses for Primate Origins

1. Choose one of the hypotheses for primate origins. From your understanding of evolution at this point in the course, why do you think your choice (hypothesis) is weak or strong?

2. In what geographical regions of the world were proprimates and euprimates found? In what geological epoch (include dates) are they placed? Why are proprimates questionable candidates as the first true primates?

3. Euprimates are placed in two closely related groups. a) What are these two groups______________ and______________; b) The two groups of euprimates are most similar to modern ______________ and______________.

4. Name the first anthropoids and list the geological dates (include the epoch) and region in which they were found. What are the key traits of these fossils? Do their locations have any implication for the origins of the higher primates?

Section 5.2: Anthropoids: The Higher Primates

1. Define catarrhines and platyrrhines.

2. List the names, geological epoch, and dates of the three groups of higher primates. Including the dates of divergence, which groups might be ancestral to platyrrhines and catarrhines?

3. What is *Branisella boliviana*? How did anthropoids get to the New World? Discuss the various hypotheses detailed in the text.

4. What is *Aegyptopithecus zeuxis*? List the date, geological epoch, and four traits of *Aegyptopithecus zeuxis*. Why is *Aegyptopithecus zeuxis* important?

5. What group do the proconsulids represent? List the date, geological epoch, and four traits of the proconsulids. Why are they important?

Section 5.3: Dryopithecines: The Mid-Miocene Apes

1. What group do the dryopithecines represent? List the date, geological epoch, and four traits of the dryopithecines. Why are they important?

CHAPTER 6

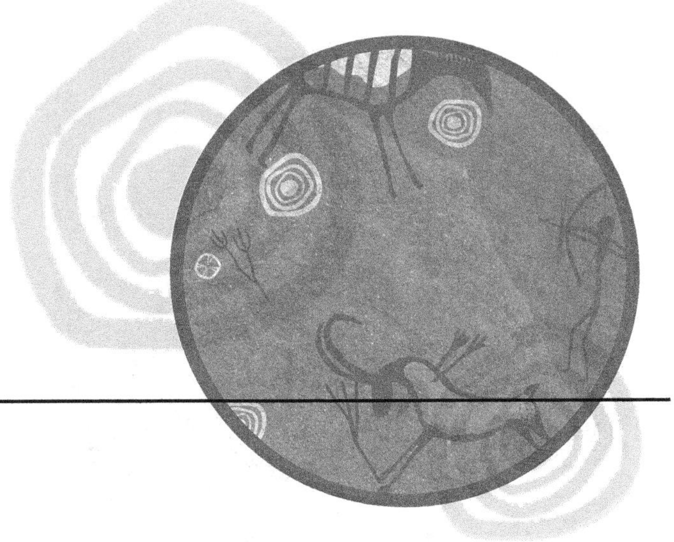

Living Primates

Biogeography of Primates

Most of the nonhuman primates are generally distributed within the tropical regions of the world, specifically between the Tropic of Cancer and Tropic of Capricorn (at this point, I will simply use the word *primates* when referring to the nonhuman primates). Above and below these two latitudes, where the regions are cooler, there are very few primates. This pattern of distribution is no accident. In the evolution of primates, the *oldest* fossils are found in the hot, dry regions of Africa. In short, primates (humans included) are tropical animals.

So, what is a **primate**? According to Nystrom and Ashmore (2008), it is "an order in the class Mammalia defined by a suite of anatomical and behavioral traits including being *large-brained*, mostly *tree-dwelling mammals*, that are *highly social*" (p. 397, italics my emphasis). The emphasis in the quote above is important for several reasons: (1) compared to the average mammal, primates have a larger brain size relative to their body size; (2) primates have traits that evolved for adaptation to life in the trees; and (3) social animals, on average, have the largest relative brain size compared to solitary animals. Let us now discuss the general characteristics (or adaptations) that are found within the order Primates.

General Primate Characteristics (Adaptations)

1. **Generalized skeleton**: primates have a flexible skeleton that enable them to grab hold, climb, and swing from tree to tree. The five long hand and foot phalanges enable primates to grab hold of the branches as they climb and to pick fruits and leaves (Figures 6.1a,b). The thumb (first digit) and index finger (second digit) enhance precision grip (Figure 6.2). The apes (**gibbons**, orangutans, **gorillas**, and **chimpanzees**) have some precision grip ability, but humans have the best precision grip ability. Also, humans have grasping hands but not grasping feet (Figure 6.3a,b). Form follows function; in human evolution the foot phalanges have evolved to be short and stubby to facilitate **bipedal** locomotion.
2. **Enhanced sense of touch**: primates have fingernails that protect the sensitive pads at the end of the fingers and toes. These sensitive pads enable primates to detect what is being held. There are also prints, commonly known as fingerprints or toe prints, on the sensitive pads, which provides a better grip on the surface the primate is moving upon.
3. **Enhanced sense of vision**: primates' primary sense (particularly monkeys, apes, and humans) is vision. We have forward-facing eyes; **stereoscopic-binocular vision** that provides overlapping fields of vision producing distance and depth perception; and color vision. For instance, forward-facing eyes enable the primate to hunt and catch insects in an **arboreal** (tree-living) environment. The stereoscopic-binocular vision enables the primate to survive (and not fall) while climbing and swinging from branch to branch. Finally, color vision enables the primate to know when leaves, flowers, and fruits are ready to be eaten. In short, these traits have evolved for life in the trees.

(a)

(b)

Image © Morgan Lane Photography, 2010. Used under license from Shutterstock, Inc.

Figure 6.1a,b. A squirrel monkey with firm hold of a tree branch (a) and a sketch of a monkey grasping and eating a banana (b)
Image © Peter Wollinga, 2010. Used under license from Shutterstock, Inc.

4. **Reduced sense of smell**: most primates (particularly monkeys and apes) do most of their activities during the day. They are said to be **diurnal**. This explains the enhanced sense of vision and the reduction of the sense of smell. The reduction in the sense of smell results in the change from a snout or **rhinarium**, like a dog, to a flat face. An enhanced sense of smell is adaptive during nighttime (**nocturnal**) activities when vision is hampered by darkness. To counter this, these animals have rhinariums or moist noses to detect scents in their habitat. Only the most primitive

Figure 6.2. Precision grip (human).
Image © ilker canikigli, 2010. Used under license from Shutterstock, Inc.

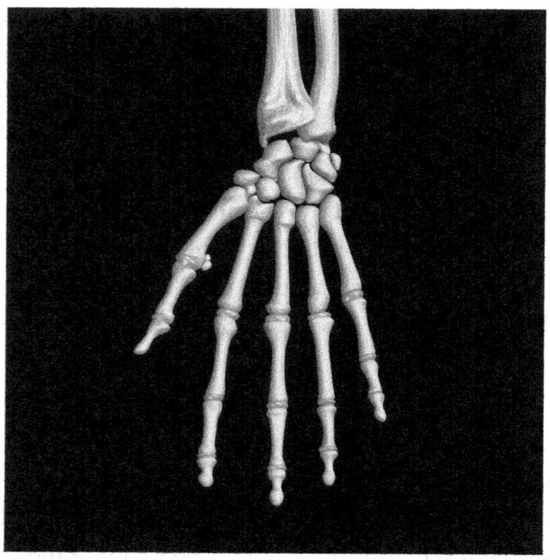

(a)

Figure 6.3a,b. Humans have grasping hands (a), but not grasping feet (b) because of bipedal locomotion.
Image © Aaliya Landholt, 2010. Used under license from Shutterstock, Inc.

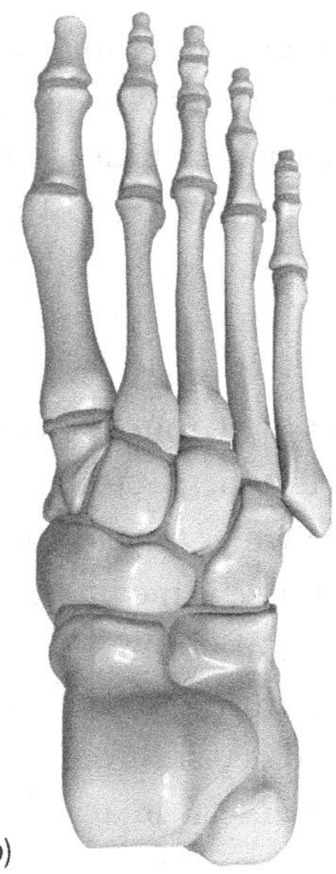

(b)

Image © Sebastian Kaulitzki, 2010. Used under license from Shutterstock, Inc.

primates have a rhinarium and, therefore, have a heightened sense of smell (Figure 6.4). They, in fact, have retained a primitive mammalian trait.

5. **Dietary flexibility**: primates are **heterodontic**, meaning that they have different types of teeth: incisors for biting and gripping; canines for tearing; premolars and molars for pulverizing and grinding food. Heterodontism enables primates to eat many different types of food, and thus ensures survival when their staple or favored food is scarce. Despite the shared heterodontism in prosimians, monkeys, and apes, they do have different cusp patterns, which reflect the differences in *how* and *what* they eat. For instance, many prosimians have a dental comb where the lower incisors and canines are elongated and packed tightly together like a comb. They use it to scrap gum (a thick, sticky fluid high in protein produced when a tree is damaged) from tree trunks. They also have a raised molar cusps pattern in order to impale insects during chewing. In contrast, monkeys have a ridged molar cusps pattern in order to process seeds and leaves. Apes, like humans, have flat molar cusps for crushing foods. Furthermore, monkeys and apes have very large canines. These canines are not only important for tearing (as mentioned above) but for creating a cutting edge in the lower teeth row. Specifically, the back of the upper canine touches the front of the lower premolar, forming a sharp cutting edge for slicing foods. For this slicing mechanism to work, the primate must be able to open and close its jaws. To facilitate this function, there are spaces (**diastema**) in the upper and lower jaws for the canines to fit when the jaws are shut.
6. **Parental investment**: primate females give birth to very immature and few offspring. Consequently, infants depend on their mothers for a very long time (Figure 6.5). During this dependency, infants learn all about their culture from their mothers in order to survive to adulthood. For example, infants learn how to make tools, obtain food, and get along within the group. This long infant dependency is strongly correlated with intelligence because, compared to the average mammal, primates have the largest relative brain size. In summary, primates are social animals; they are prone to forming social groups; and they need the touch or "warmth" of other bodies to reinforce social bonds (Figure 6.6). This is why **grooming** (picking dry skin and insects from another individual's fur) is so important in strengthening bonds and reducing tension within the group. Finally, most fathers are not involved in infant care.

Figure 6.4. Lemurs, one of the most primitive primates, have a rhinarium which give them an enhanced sense of smell.

Figure 6.5. An old world monkey mother holds her infant.

Figure 6.6. Two baboons "kissing," which is a distinctive type of grooming behavior.

Taxonomic Classification

Taxonomy is the classification of plants and animals into categories based on similarities and differences. The classification scheme listed below was originated by Linnaeus (see Chapter 2). It becomes very inclusive (more biological life forms) as you move up the taxonomy, and less inclusive or more *specific* as you move down the taxonomy. At this level of the taxonomy, the animal has two Latin names (**binomen**). In general, Primates is an *order* that is a subdivision of the *class* Mammalia that is within the *phylum* Chordata. In this text, the traditional primate taxonomic classification will be used.

(a)

Image © liveo, 2010. Used under license from Shutterstock, Inc.

Modern Taxonomic Classification:

	Kingdom (Animalia)
Inclusive	**Phylum** (Chordata—animals with a backbone)
↑	**Class** (Mammalia—animals with mammary glands)
	Order (Primates)
	Suborder
	Infraorder
	Superfamily
↓	**Family**
Exclusive	**Genus**
	Species

(b)

Image © Thierry Dagnelie, 2010. Used under license from Shutterstock, Inc.

PROSIMIANS (STREPSIRHINI—MOIST NOISE)

Order: Primates
Suborder: Prosimii

Some examples of **prosimians** are lemurs, tarsiers, sifakas, lorises, and bush babies (Figure 6.7a,b,c). Most of these prosimians have the following respective characteristics and behaviors: (1) they have moist noses and therefore a well-developed sense of smell; (2) they are adapted to nocturnal activities; (3) they have claws on some digits; (4) they are vertical clingers and leapers; and (5) they forage alone (solitary) but frequently form sleeping groups.

ANTHROPOIDEA (HAPLORHINI—LACK OF MOIST NOSE)

Order:	**Primates**
Suborder:	*Anthropoidea* (Haplorhini—(lack moist noses)
Infraorder:	*Catarrhini* & *Platyrrhini*

The suborder Anthropoidea includes all monkeys, apes, and humans. The terms **Catarrhini** and **Platyrrhini** refer to the nose. Catarrhines have downward-facing nostrils while platyrrhines have laterally facing nostrils. Catarrhines include all **Old World monkeys**, apes, and humans. Platyrrhines include all **New World monkeys**.

(c)

Figure 6.7a,b,c. Tarsier (a); Lemur (b); and sifaka (c).

Image © Eric Isselée, 2010. Used under license from Shutterstock, Inc.

Order: **Primates**
Suborder: *Anthropoidea*
Infraorder: *Catarrhini* & *Platyrrhini*

Superfamily:
-Hominoidea &
Cercopithecoidea

(infraorder: ****Platyrrhini***)
Superfamily:
-Ceboidea

After the infraorders Catarrhini and Platyrrhini comes the superfamilies Hominoidea and Cercopithecoidea (under Catarrhini) and Ceboidea (under Platyrrhini). **Hominoidea** includes all apes and humans. **Cercopithecoidea** includes all Old World monkeys. **Ceboidea** includes all New World monkeys.

Family

Cercopithecidae	***Hominidae***	***Cebidae***
Colobinae	***Pongidae***	***Callitrichidae***
	Hylobatidae	***Atelidae***

Genus:	***Hylobates***	***Pongo***	***Gorilla***	***Pan***	***Homo***
Species:	***lar***	***pygmaeus***	***gorilla***	***troglodytes***	***sapiens***
				Pan	
				paniscus	

After the superfamilies Hominoidea, Cercopithecoidea, and Ceboidea come the families Cercopithecidae, Colobinae, Hominidae, Pongidae, Hylobatidae, Cebidae, and Callitrichidae. Cercopithecidae and **Colobinae** are different families of Old World monkeys. **Cebidae** and Callitrichidae are two families of New World monkeys. **Hylobatidae** includes gibbon species (lesser ape). **Pongidae** includes all the great apes. **Hominidae** includes all humans. After the family group comes the genus and species. The major genus and species groups are *Hylobates lar* (gibbon); *Pongo pygmaeus* (orangutan); *Gorilla gorilla* (western lowland gorilla); *Pan troglodytes* (common chimpanzee); *Pan paniscus* (pygmy chimpanzee [bonobo]); and *Homo sapiens* (humans). Except for humans, there are two or more species in gibbons, orangutans, gorillas, and chimpanzees. These other species will not be dealt with in this text. Now let us take each primate group separately and study their characteristics.

New World Monkeys (Ceboidea)

They have several key characteristics (see Figure 6.8a,b,c):

1. Laterally facing nostrils
2. **Prehensile** (grasping) **tail**
3. Total premolars: 12
4. Total teeth: 36
5. Dental formula: 2-1-3-3 (one side of the mouth—upper or lower/left or right)
6. Diurnal (99 percent)
7. Mostly arboreal habitual quadrupeds
8. All limbs the same length
9. Smaller than Old World monkeys
10. Diet: fruit, leaves, gum, insects, small mammals

(a)

Figure 6.8a,b,c. Cacajao (a); Howler monkey walking on all fours on the tree branches (arboreal quadrupedalism) (b, c).

(b)

(c)

11. Social organization: variable
12. Low frequency of **sexual dimorphism** (body size variability in males and females)

Primate social organization is defined as the way in which primate groups organize themselves, such as single male/single female; multimale/multifemale; single male/multifemale; single female/multimale; or a family group (male, female, and offspring). New World monkey social organizations vary within and between groups. Interestingly, some fathers are involved in infant care (Figure 6.9).

Furthermore, some species of New World monkeys have special guts to digest mature leaves. Mature leaves are difficult to digest because of the cellulose that is a part of their composition. Some New World monkeys such as the howler and the muriqui have **sacculated stomachs** containing special enzymes to break down the cellulose in mature leaves, enabling them to make use of this abundant resource. Finally, there is one nocturnal monkey, which is the owl monkey.

Old World Monkeys (Cercopithecoidea)

1. Downward-facing nostrils
2. No prehensile (grasping) tail
3. Total premolars: 8
4. Total teeth: 32
5. Teeth: molar cusps: **bilophodont**
6. Dental formula: 2-1-2-3 (one side of the mouth—upper or lower/left or right)
7. Diurnal (100 percent)
8. Arboreal and **terrestrial** (on the ground) habitual quadrupeds
9. All limbs the same length
10. Larger than New World monkeys
11. Diet: fruit, leaves, gum, insects, small mammals
12. Social structure: variable
13. Males are dominant (over females)
14. High frequency of sexual dimorphism (body size variability in males and females)

Figure 6.9. In squirrel monkeys, the female is the primary caretaker of her offspring, but males do occasionally carry the infant.

(a)

Image © Kjersti Joergensen, 2010. Used under license from Shutterstock, Inc.

(b)

Image © Eric Isselée, 2010. Used under license from Shutterstock, Inc.

(c)

Image © VanHart, 2010. Used under license from Shutterstock, Inc.

(d)

Image © EcoPrint, 2010. Used under license from Shutterstock, Inc.

(e)

(f)

Image © JinYoung Lee, 2010. Used under license from Shutterstock, Inc.

Figure 6.10a,b,c,d,e,f. Proboscis monkey (a); Guenons (b); Mandrills (c); Rhesus macaque (d); Hamadryas baboon (e); and Japanese macaque (f).

Image © wollemipine, 2010. Used under license from Shutterstock, Inc.

Old World monkeys are a very diverse group of primates. They are able to adapt and survive in many different types of environments, such as tropical rainforest, tropical seasonal forest, tropical woodland forest, tropical savanna, and semidesert regions (Figure 6.10a,b,c,d,e,f). Related to this is the fact that some species of Old World monkeys (Colobinae), like the howler and muriqui, have a special gut to digest mature leaves. This adaptation certainly expands their ranges not only in Africa but also in Asia. Finally, there is male-male competition for females. This explains the high frequency of sexual dimorphism—males are twice as large as females in body size and canine size.

Apes (*Hominoidea*)

General characteristics:

1. No tails
2. Larger than all monkeys
3. Brains are larger and more complex than those of all monkeys
4. Dental formula: 2-1-2-3 (one side of the mouth—upper or lower/left or right)
5. **Y-5 molar cusps pattern** (between the five cusps, the grooves form a "Y" pattern)
6. More parental investment than all monkeys
7. Diet: fruit, immature leaves, flowers, insects
8. Social structure: variable
9. High degree of sexual dimorphism (body size variability in males and females) in some species
10. Larger, stronger collar bone than all monkeys*
11. Shoulder joint is very flexible and capable of a wide angle of movement*
12. Shoulder blades are located more toward the back*

    *Enable apes to raise their arms over their heads and hang or swing
13. Longer arms than legs (humans: longer legs than arms)
14. Longer fingers than all monkeys

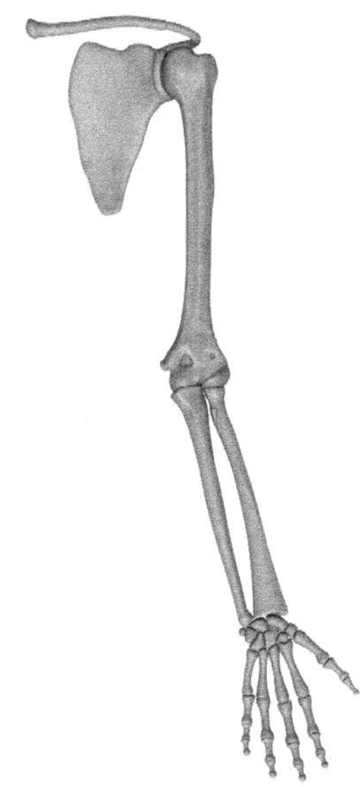

Figure 6.11. The shoulder joint of hominoidea that facilitates hanging and swing.

Image © 3drenderings, 2010. Used under license from Shutterstock, Inc.

There are two categories of apes from two regions of the Old World: (1) the lesser apes (Southeast Asian gibbons); and (2) the great apes (Southeast Asian orangutans and African apes: gorillas and chimpanzees). Some are proficient swingers and others are expert climbers (Figure 6.11). Let us now discuss each ape species separately.

Gibbon (*Hylobates lar*)

1. Smallest of the living apes
2. Adapted to life in the trees (highly agile movement: brachiation)
3. Longest arms of the living apes
4. Fingers are long/thumb short
5. No sexual dimorphism
6. Social organization: family group (male, female, and offspring)
7. Diet: fruit and immature leaves

Figure 6.12. White-cheeked gibbon.

Image © Nagel Photography, 2010. Used under license from Shutterstock, Inc.

Since there are no size differences between males and females, it would be very difficult to distinguish the sexes. But this is not the case. Evolution has found a way to differentiate them, and this is by a different-color pelage (coat), called **sexual dichromatism**. Family groups mate for life; therefore, there is no selection for males to be twice as large as females (Figure 6.12).

Orangutan (*Pongo pygmaeus*)

1. Larger than gibbons (smaller than gorillas)
2. Adapted to life in the trees (**quadramous**: using all four limbs to vault from tree to tree—adult males do travel on the ground)

3. **Quadrupedal** fist walkers on the ground
4. High degree of sexual dimorphism (males twice as large as females)
5. Adult males have large fat pads on their faces (to increase the resonance of vocal communication)
6. Diet: fruits and immature leaves
7. Social organization: solitary—males and females have separate lives
8. Males are dominant (over females)
9. Birth spacing (time between birth of first infant and second infant): seven years

The orangutan males and females live solitary lives. Why? It could be that there is not enough food in one area to support males, females, and offspring. Nonetheless, the alpha male has his territory in which several females (polygynous—harem) and their offspring reside. The alpha male roams his territory in search of food and occasionally seeks out the females in order to mate. Orangutan females have no distinct mating cycle or estrus like monkeys and chimpanzees. In fact, female orangutans have similar gestation lengths (8–9 months) as in gorilla and human females. Furthermore, primate females in general and orangutan females in particular are not able to have lactation and ovulation simultaneously (like human females). In other words, when the female is lactating there is no ovulation and vice versa. As a result, birth spacing is very wide in primates (Figure 6.13a,b,c).

Gorilla (*Gorilla gorilla*)

1. Largest of the living apes
2. Spend most of the time on the ground because of their large size
3. Quadrupedal **knuckle walkers** on the ground
4. High degree of sexual dimorphism (males twice as large as females)
5. Adult males have a sagittal crest (a ridge of bone in the center of the skull for the attachment of chewing muscles)
6. Fully adult males have grey hair on their backs ("silverbacks")
7. Diet: mostly fruit and leaves
8. Social organization: polygynous—single male/multifemale
9. Males are dominant (over females)
10. Birth spacing (time between birth of first infant and second infant): five years

(a)

(b)

(c)

Figure 6.13a,b,c. Orangutan male with the characteristic fat pads on face (a); orangutan moving along the rope using all four limbs (quadramous)(b); and orangutan mother with infant (c).

The alpha male or silverback is very protective of his harem (Figure 6.14). He is intolerant of other adult males. But he will tolerate a subadult male because this subadult male does not have the physical tools, that is, large canines and body size, to challenge the alpha male and, therefore, will not mate. Also, he might be the alpha male's son. Like the orangutan females, gorilla females have similar gestation lengths (nine months) as in human females. Birth spacing is wide (five years), but not as wide as in orangutans. The silverback is in charge so that when he moves the females and offspring move with him (Figure 6.15a,b).

Figure 6.14. Silverback male gorilla.

Chimpanzee 1 (*Pan troglodytes*)—common chimpanzee

1. Spend most of the time on the ground and in the trees
2. Quadrupedal knuckle walkers on the ground
3. Slight sexual dimorphism (males slightly larger than females)
4. Diet: fruits, leaves, seeds, nuts, insects, and meat
5. Social organization: multimale/multifemale
6. Males are dominant (over females)
7. Strongest social bonds between males
8. Birthing: females have an **estrus** cycle (visible sign when ready to mate)
9. Tool use observed: rocks used to break open nuts; sticks used to "fish" for termites; leaves used as a sponge to obtain water

(a) (b)

Figure 6.15a,b. Female gorilla (a) and sketch of gorilla in a quadrupedal knuckle-walking stance (b).

(a)

Figure 6.16a,b. Sketches of common chimpanzee in normal quadrupedal knuckle-walking stance (a) and modified bipedal stance (b).

Image © Eric Isselée, 2010. Used under license from Shutterstock, Inc.

(b)

Image © Eric Isselée, 2010. Used under license from Shutterstock, Inc.

Among the apes, both species of chimpanzees are very intelligent (Figure 6.16a,b). The common chimpanzee has been observed using tools: "fishing" for termites and hunting monkeys. Behaviors such as finding a tree branch and modifying it so that it is the right width and length to fit into a termite hole, or several chimpanzees cooperating in order to force a prey (the monkey) into the "trap" involve using sophisticated cognitive processes (Figure 6.17). Related to this cooperation is male **dominance**. Unlike the rigid gorilla and orangutan social organizations, the dominant male chimpanzee must rely on alliances (friends and family) in addition to strength and intelligence to maintain his position. Finally, males are generally dominant over females. But high-ranking females are dominant over low-ranking males. And the mother infant bond is very strong (Figure 6.18).

Figure 6.17. Common chimpanzee and tool-use.

Image © Norma Cornes, 2010. Used under license from Shutterstock, Inc.

Figure 6.18. Chimpanzee mother holding her infant.

Image © Knud Nielsen, 2010. Used under license from Shutterstock, Inc.

Figure 6.19a,b. Pygmy chimpanzee or commonly known as the bonobo.

(b)

Chimpanzee 2 (*Pan paniscus*)—pygmy chimpanzee **(bonobo)**

10. Spend most of the time on the ground and in the trees
11. Quadrupedal knuckle walkers on the ground
12. Smaller than the common chimpanzee
13. Slight sexual dimorphism (males slightly larger than females)
14. Diet: fruits, leaves, seeds, nuts, insects, and meat
15. Social organization: multimale/multifemale
16. Females are dominant (over males)
17. Strongest social bonds between females
18. Birthing: females have an estrus cycle (visible sign when ready to mate)
19. Tool-use observed: rocks used to break open nuts; sticks used to "fish" for termites; leaves used as a sponge to obtain water

The bonobo's (pygmy chimpanzee) social organization is quite different compared to that of the common chimpanzee (Figure 6.19a,b). First, females are dominant among bonobos. This relates to the fact that the strongest bonds are between females. In essence, the dominance status of a male depends on his mother. Second, mating is done face-to-face unlike the traditional mounting position observed in most primate species. In addition, sexual activity (in addition to grooming) is used frequently to reduce tension and avoid conflict. This is unusual because grooming is used exclusively by all primate groups to reduce tension and avoid conflict.

In summary, primates are biologically and culturally complex (Figure 6.20a,b,c,d). Their complex behaviors make them ideal living models in uncovering human origins. Behaviors such as tool use and cooperative hunting in chimpanzees, discussed earlier in the chapter, are well established based on field studies. But more controversial behaviors, such as language acquisition, male-female reproductive strategies, and **altruistic** behaviors (kin selection), are still being debated by scientists. Primates, like any other animal, use vocal and nonvocal (gestures) communication (Figure 6.21). But are gorillas and chimpanzees capable of language acquisition (using symbols)? Many field studies have been done (Gardner and Gardner 1969; Patterson 1978; Patterson and Linden 1981; Rumbaugh 1977; Savage-Rumbaugh and Lewin 1994; Terrace 1987), but the debate continues. Finally, would individual primates frequently put themselves in danger for the good of their kin (rather than strangers)? This is very difficult to observe in the wild since the researcher must first determine if individuals within the group are related.

(a)

Image © Eric Gevaert, 2010. Used under license from Shutterstock, Inc.

(b)

Image © Rob Schumann, 2010. Used under license from Shutterstock, Inc.

(c)

(d)

Image © Ajancsco, 2010. Used under license from Shutterstock, Inc.

Figure 6.20a,b,c,d. Gorilla male and female touching (a); Guenon (b); Lion tamarin (c); and chimpanzee mother with infant and juveniles (d).

Image © Eric Gevaert, 2010. Used under license from Shutterstock, Inc.

Figure 6.21. Depictions of chimpanzee gestures and vocalizations.

Image © Eric Isselée, 2010. Used under license from Shutterstock, Inc.

Summary of Key Ideas

1. Primate social organization is the way primate groups organize themselves into single male/single female; multimale/multifemale; single male/multifemale; single female/multimale; or a family group (male, female, and offspring).
2. Most primate species are distributed within the tropical regions of the world.
3. Primates are large-brained, tree dwelling, and highly social mammals.
4. Primates have a generalized skeleton, enhanced sense of touch, enhanced vision, reduced sense of smell, dietary flexibility, and a high degree of parental investment.
5. Primates have different types of teeth: heterodontic.
6. Primates' varied molar patterns reflect variation in diet.
7. Primate females give birth to very immature and few offspring, indicating a long infant dependency compared to other mammals.
8. The mother-infant bond is very strong; most fathers are not involved in infant care.
9. Estrus is a visible sign when the female (e.g., chimpanzee) is ready to mate.
10. Taxonomy is the classification of plants and animals based on similarities and differences: prosimians, monkeys, apes, and humans are in the class Mammalia and order Primates.
11. Prosimians are the most primitive primates. They have a primitive mammalian trait, which is a heightened sense of smell (a moist nose).
12. The suborder Anthropoidea includes all monkeys, apes, and humans.
13. The infraorders Catarrhini and Platyrrhini refer to the nose: downward facing (in catarrhines) and laterally facing (in platyrrhines).
14. Catarrhini includes all Old World monkeys, apes, and humans. Platyrrhini includes all New World monkeys.
15. The superfamily Hominoidea includes apes and humans. The superfamily Cercopithecoidea includes all Old World monkeys. The superfamily Ceboidea includes all New World monkeys.
16. Cercopithecidae, Colobinae, Cebidae, Hylobatidae, Pongidae, and Hominidae are all family groups. Cercopithecidae and Colobinae are two Old World monkey groups. Cebidae includes all New World monkeys; Hylobatidae includes all gibbons; Pongidae includes all great apes; and Hominidae includes all humans.
17. New World monkeys have a total of 36 teeth. Within that total, they have 12 premolars for a dental formula of 2-1-3-3. Some of them have prehensile (grasping) tails and all are mostly arboreal.
18. Old World monkeys have a total of 32 teeth (like apes and humans). Within that total, they have eight premolars for a dental formula of 2-1-2-3. They are both terrestrial and arboreal.
19. Some New and Old World monkey species have a special adaptation (sacculated stomach) to digest mature leaves.
20. Within the suborder Anthropoidea, primate groups conduct activities during the daytime (diurnal). However, there is one nocturnal New World monkey: the owl monkey.
21. Sexual dimorphism is size differences between males and females within any mammalian species. When males are twice as large as females in primate species, this indicates male-male competition for females.
22. Gibbon males and females are the same size. This indicates a monogamous relationship; no male-male competition for females.
23. In monkeys, all limbs are the same size. In apes, the arms are longer than the legs.
24. Old World monkeys have a bilophodont molar cusps pattern; apes and humans have a Y-5 molar cusps pattern.
25. Gorillas and orangutans have no estrus cycle. They have a similar gestation length as human females.
26. Chimpanzees are capable of tool use and cooperative hunting.
27. In the multimale/multifemale social organization, dominance depends on intelligence, strength, and, most of all, reliance on friends and alliances.

28. In most primate societies, males are dominant. But within bonobo society, it is females who are dominant.
29. Grooming is an essential behavior in primate societies. It reinforces social bonds and reduces conflict.

Key Terms

Altruism
Anthropoidea
Arboreal
Bilophodont molars
Binomen
Biogeography
Bipedalism
Bonobo
Catarrhini
Cebidae
Ceboidea
Cercopithecoidea
Chimpanzee
Colobinae
Diastema
Diurnal
Dominance
Estrus
Gibbon
Gorilla
Grooming
Haplorhini
Heterodontic
Hominidae
Hominoidea
Hylobatidae
Knuckle walking
New World monkey
Nocturnal
Old World monkey
Platyrrhini
Pongidae
Prehensile tail
Primates
Prosimian
Quadramous
Quadrupedalism
Rhinarium
Sacculated stomach
Sexual dichromatism
Sexual dimorphism
Stereoscopic-binocular vision
Strepsirhini
Taxonomy
Terrestrial
Y-5 molar pattern

Further Readings

De Waal, F. 1997 *Bonobo: The forgotten ape.* Berkeley: University of California Press.
Falk, D. 2000. *Primate diversity.* New York: W. W. Norton.
Goodall, J. 1986. *The chimpanzees of Gombe: Patterns of behavior.* Cambridge, MA: Harvard University Press.
McGrew, W. C. 1998. Culture in nonhuman primates. *Annual Review of Anthropology* 27: 301–28.
Strier, K. 2007. *Primate behavioral ecology,* 3rd ed. Boston: Allyn & Bacon.

INTERNET RESOURCES

Primate Info Net: About the Primates
http://pin.primate.wisc.edu/aboutp
This Web site is maintained by the Wisconsin Regional Primate Research Center. It provides extensive information about primates.

African Ape Study Sites
http://weber.ucsd.edu/~jmoore/apesites/index.html
This Web site provides information on sites where African apes (the two species of chimpanzees and the gorillas) have been and are now being studied.

Gombe Chimpanzee Culture
http://www.janegoodall.org/chimp_central/default.asp
This is pioneer primate researcher Jane Goodall's Web site describing chimpanzee behavior and conservation.

Chapter 6: Living Primates

Study guide: By knowing the material below you will be well prepared for the exam. Don't forget to read the chapter and study your notes (this includes the chapter worksheets).

1. Know the primate taxonomy
2. Know the characteristics of Old World Monkeys (OWM), New World Monkeys (NWM), and apes
3. Know primate arboreal adaptations
4. What is the most common type of social group found in primate societies?
5. What is the structure of a polygynous primate social group?
6. Be able to define dominance hierarchy
7. Understand parental investment
8. Be able to explain why a new dominant male baboon might kill infant (suckling) baboons. What is this called?
9. The most biologically primitive of living primates are__________. Know their characteristics
10. Be able to list at least two (2) prosimians.
11. Anthropoids consist of____________. Where do they live (geographically)?
12. Monkeys are found in the two geographical regions generally known as_________ and________.
13. Know the dental formula of OWMs and NWMs
14. What is a prehensile tail? Which monkeys have this trait: OWM or NWM?
15. Be able to define arboreal
16. Be able to define terrestrial
17. Are Howler monkeys NWM or OWM?
18. Howler and Colobus monkeys have a special adapation. What is it and what is the purpose of this adaptation?
19. Which group of primates is closer (genetically) to humans: Prosimians, NWM, OWM, or Apes?
20. Which primate species is closer (genetically) to humans: lemurs, howler monkeys, baboons, orangutans, gorillas, or chimpanzees?
21. Are baboons NWM or OWM?
22. Hominoids consists of___________.
23. Geographically, where do one find apes?
24. The "Y-5" molar pattern is found in living___________.
25. Generally, which primate group has longer arms than legs?
26. Which primate species has the longest arms?
27. The African apes consist of__________.
28. The Asian apes consist of___________.
29. DNA sequence between chimpanzees and humans are______% identical.
30. Define sexual dimorphism. Be able to name two ape species where sexual dimorphism is most noticeable. What does it suggests about social organization (mate selection)?
31. How does a large adult male orangutan get from one tree to another?
32. In primate societies, what is a "silverback"?
33. What is the diet of gorillas?
34. In bonobo (pygmy chimpanzee) society__________tend to be the most dominant.
35. In addition to grooming, bonobos reduce tension in another way. What do they do?
36. Which OWM group is sexually dimorphic?
37. Which ape species is sexually monomorphic?
38. Describe the cusp pattern of bilophodont molars. In which primate species do you find them: Monkeys, apes, or humans?
39. What is grooming? How is this behavior important in primate societies?

NAME ______________________________ **DATE** ______________

Chapter 6: Living Primates

HOME WORK EXERCISES (10PTS.)

The goal of the home work exercises is assess your knowledge of the concepts and ideas presented in the chapter. Please answer the questions in each section as you go. When class lecture on this chapter is completed, carefully tear the home work exercises from the textbook and turn it in. **[Bonus pts. = 10]**

Use a pen or pencil to answer the questions below. Make sure your writing is LEGIBLE.

Section 6.1: Biogeography of Primates

1. After reading the first paragraph of this chapter, explain 'biogeography.'

2. What is a primate?

Section 6.2: General Primate Characteristics (Adaptations)

1. List the general primate characteristics. Why do primates have stereoscopic binocular vision?

2. Why do monkeys and apes have an enhanced sense or vision and a reduced sense of smell?

3. Generally, primate females invest more in their offspring than do many other mammals. What is the long-term impact of this investment on the offspring?

4. Discuss the advantages of heterodontism in primates.

Section 6.3: Taxonomic Classification

1. List the taxonomic categories? What is binomen?

2. What are the two suborders in primate taxonomy?

3. Catarrhini and Platyrrhini are_______________in primate taxonomy. They refer to the__________. Name two primate groups in Platyrrhini and two primate groups in Catarrhini.

4. Hominoidea, Cercopithecoidea, and Ceboidea are_____________________in primate taxonomy. Name the primate groups in Hominoidea, Cercopithecoidea, and Ceboidea.

5. List all the family groups in primate taxonomy. Where do Old World monkeys, New World monkeys, apes, and humans fit?

6. Looking at the list of traits for New World monkey, what is the one key trait that makes them different from Old World monkeys? Explain.

7. Define sexual dimorphism. When male primates are twice as large as female primates in body size and canine size, what does this tell us about the social structure of the group?

8. In the Old and New World, a few monkey species have sacculated stomachs. Name the monkey species. Explain this adaptation.

9. What is dental formula? What is cusps pattern? Now, make a table with New World monkeys, Old World monkeys, and apes. List the dental formula and cusps pattern for each.

10. List the Asian apes and the African apes. Indicate the lesser ape and the great apes.

11. Make a table with monkeys, gibbons, orangutans, and gorillas. List the locomotion (how they move in the trees and/or on the ground) for each. Generally, what is the arm-to-leg length differences of each primate?

12. When male and female primates are the same size, what does this tell us about the social structure of the group? Give one primate example. How does evolution distinguish them?

13. Monkeys and chimpanzees have estrus cycles: a visible sign when the female is ready to mate. Do orangutans and gorillas have this same reproductive feature? Explain.

14. Adult male gorillas have a sagittal crest. What/where is this feature on the body and what is its function?

15. A silverback male is a____

16. Why does a male gorilla tolerate subadult males (and not other adult males)?

17. Choose the correct answer and write it in the space provided: females/aggressive; females/non-aggressive; males/aggressive; males/non-aggressive
   a) In *Pan troglodytes*______________are dominant and________________.
   b) In *Pan paniscus*________________are dominant__________________.

18. Other than grooming, what other behaviors are used by bonobos to reduce tension after conflict?

19. Define altruism and kin selection. Keeping primate relationships in mind, in what type of group is altruism mostly found?

CHAPTER 7

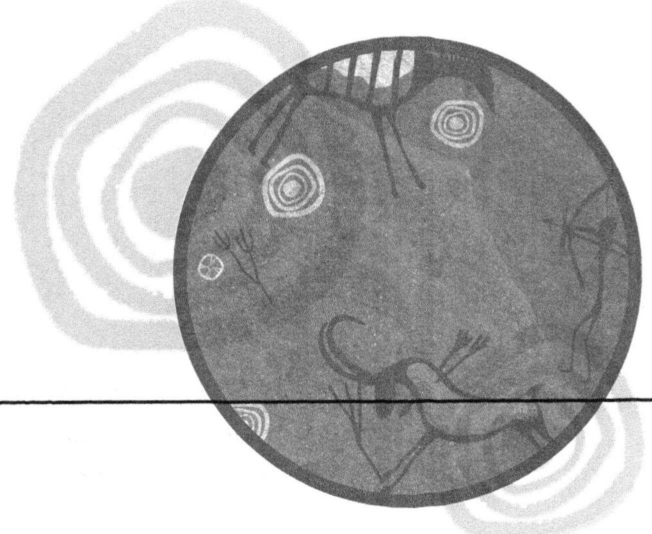

Dating Methods

Geological Time Period

The earth is approximately 4 billion years old. To cope with this immense age, geologists have organized the earth's history into divisions and subdivisions. For instance, there are two *eons* (Precambrian and Phanerozoic), each of which are major divisions. Each of these eons is broken into several subdivisions called *eras*. Finally, the eras are further divided into *periods*. During the **Precambrian eon**, dating from 4 billion years ago until 545 million years ago, single-celled and simple multicelled organisms first evolved (Relethford 2005). During the **Phanerozoic eon**, from 545 million years ago to the present, invertebrates and vertebrates (fish, amphibians, reptiles, dinosaurs, mammals, and primates) evolved. The Phanerozoic eon is divided into three geological eras: **Paleozoic** (545–245 million years ago), **Mesozoic** (245–65 million years ago), and **Cenozoic** (65 million years ago to the present) (Larsen 2008; Relethford 2005).

PALEOZOIC ERA (545–245 MILLION YEARS AGO)

1. Origin of invertebrates
2. Origin of vertebrates:
   a. Fish and land plants
   b. Amphibians and forests
   c. Reptiles and insects
   d. Mammal-like reptiles

By the late Paleozoic, all the continents were joined together in a hypothetical landmass called **Pangaea** (Cowen 1995) (Figure 7.1).

MESOZOIC ERA (245–65 MILLION YEARS AGO)

1. First dinosaurs and egg-laying mammals
2. First birdlike reptiles
3. First birds and placental mammals
4. Extinction of dinosaurs

The Mesozoic era is known as the "Age of Reptiles" because the reptiles, particularly the dinosaurs, were the dominant form of life for 108 million years (Figure 7.2a,b). Toward the end of the Mesozoic era, the dinosaurs became extinct. In addition, the supercontinent Pangaea began to separate.

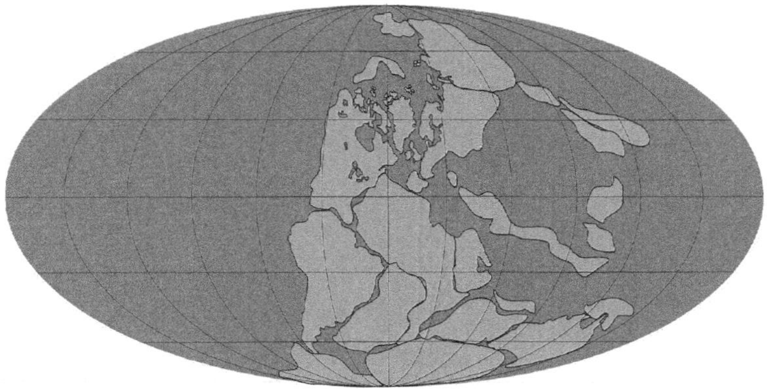

Figure 7.1. The supercontinent in the late Paleozoic called Pangaea.

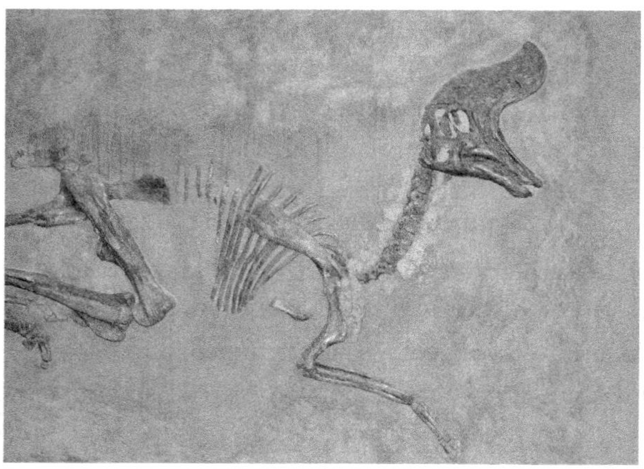

(a)

Figure 7.2a,b. Dinosaur fossils.

(b)

CENOZOIC ERA (65 MILLION YEARS TO PRESENT)

1. Diversification of modern mammals:
   a. Primates
   b. Humans

The Cenozoic era is known as the "Age of Mammals" because mammals were the dominant form of life after the dinosaurs became extinct. In essence, mammals were able to exploit the available niches vacated by the dinosaurs so that today there are thousands of genera and species of mammals. In addition, in the Cenozoic most of the continents were separate but not quite in their present positions. In fact, it is probable that the African, Antarctic, and South American continents were close enough to one another (and warm enough) to allow animals—particularly monkeys—to cross over to the South American continent. The earliest fossil evidence of monkeys in South America dates back to approximately 30 million years ago (Fleagle 1999).

When you look at a chart detailing earth's immense geological history, you will be struck by the incredibly short time period (approximately 5 million years) of human origins (see the accompanying chart). In the immense geological history of life on this planet, we humans and our earliest ancestors are very recent. It is this 5 million years that will be explored in this textbook. Let us now discuss several techniques in which dates for fossils are obtained. These techniques are critical in constructing evolutionary histories (**phylogenies**) of biological life forms.

Geological Time Period

yrs/Eras	Period	Epoch	Years BP	Main Events
Cenozoic 65 mya	**Quaternary**	**Pleistocene**	**10,000 to 1.8 mya**	Humans use/control fire
		Pliocene	**1.8 to 5 mya**	Australopithecus, Paranthropus & Homo
		Milocene	**5 to 23 mya**	First hominids & apes
		Oligocene	24 to 34 mya	Catarrhines & Platyrrhines separate
	Tertiary	Eocene Paleocene	34 to 56 mya 54 to 65 mya	First anthropoids
Mesozoic 245-65 mya	**Cretaceous**		65 to 136 mya	First flowering plants/ Extinction dinosuars
	Jurassic		136 to 190 mya	First birds/dinosaurs
	Triassic		190 to 245 mya	First mammals/Age of dinosaurs begins
Paleozoic 545-245 mya				

Dating Methods

There are two categories of dating methods: (1) **relative dating** and (2) **numerical dating.**

RELATIVE DATING

Relative dating does not give a number. It can only tell you if the object is "older" or "younger." Three examples of relative dating are stratigraphy, chemical dating, and biostratigraphy.

Stratigraphy is based on the **law of superposition**. What does this mean? As you dig deeper and deeper into the ground, you encounter different layers of sediment. These layers of sediment are called **strata** or stratigraphic layers. The layers below are older than the layers above.

Chemical dating uses the accumulation of chemicals in a fossil to determine relative age. For example, the amount of fluorine in bones buried in the strata below (strata B) should be more compared to the amount in bones buried in the strata above (strata A), simply because the bones in strata B were in the ground longer and, therefore, had more time to accumulate fluorine.

Biostratigraphy uses well-known and well-dated animal fossils to date problem sites. For instance, a site composed of limestone cannot be dated by the current numerical techniques, which require **volcanic rock**. Consequently, if you find a pig species, *Sus scrofa* (a well-dated fossil pig species found at many sites), within an assemblage of prehuman fossils, you will be able to use the dates of *Sus scrofa* to date your prehuman fossils, simply because the evolutionary history of this respective pig species is well known.

NUMERICAL DATING (RADIOMETRIC)

There are two categories of numerical dating: radiometric and nonradiometric.

Carbon 14 (**Radiometric**: based on radioactive decay). During life animals take in normal carbon-12 (stable **isotope**: 6 protons/6 neutrons) through eating plants and other animals. **Carbon-14** (unstable isotope: 6 protons/

8 neutrons), created in the atmosphere through radioactivity, is also taken in by animals. Moreover, equal ratios of carbon-12 and carbon-14 are taken in during life. When the animal dies, no more carbon is taken in. Carbon-14 begins to decay into a more stable form (nitrogen-14). The time it takes for carbon-14 to decay into a stable form is constant and can be measured mathematically. This time period is called **half-life**. For instance, it takes 5,730 years for one-half of carbon-14 to decay into nitrogen-14. This means that artifacts, such as bone, charcoal, hair, leather, pottery, soil, pollen, and wood, to name a few, 70,000–50,000 years or younger can be dated using carbon-14 (Figure 7.3). It is important to note that a portion of the object must be used (destroyed) to obtain the date.

Potassium-40/Argon-40 (Radiometric: based on radioactive decay). This dating technique works on the same principle as carbon-14. But it takes approximately 1.3 billion years for one-half of **potassium-40** to decay into **argon-40** (gas), making it possible to date fossils older than 100,000 years. Furthermore, the best material for this technique is volcanic rock. So, unlike carbon-14, it is the volcanic rock that is dated and not the fossil itself.

Figure 7.3. Fossilized wood.

Image © BW Folsom, 2010. Used under license from Shutterstock, Inc.

NUMERICAL DATING (NONRADIOMETRIC)

Thermoluminescence If you study the word **thermoluminescence** carefully, you will see that it means both "heat" (thermo) and "light" (luminescence). When artifacts such as flint or pottery are recovered from a site, the archaeologist is naturally curious about the age of these artifacts. In using this technique, the appropriate question to ask is "When was the artifact last heated?" Flint, for example, is reheated to a specific temperature until it glows or produces luminescence. This luminescence represents the trapped electrons. Therefore, the more luminescence, the more trapped electrons, or the more time that has passed since the object was last heated. This technique can be used to date fossils over 1 million years old.

Amino acid dating. All living organisms are made up of proteins. After death, protein molecules (**amino acids**) begin to decay. In other words, the amino acids modify their molecular structure, which is indicated by the direction they bend polarized light (left-handed or right-handed amino acids) when viewed under high-powered magnification (Larsen 2008). In short, the fossil date is based on the ratio of left-handed amino acid (indicating a younger fossil bone) to right-handed amino acid (indicating an older fossil bone). The longer the organism has been dead, the greater the number of right-handed amino acids.

Genetic dating. DNA technology has revolutionized the way we study evolutionary relationships among animal species. Since **genetic** differences accumulate at a constant rate, a so-called **molecular clock** can be developed to find out when closely related species have diverged. In essence, geneticists can calculate genetic distances within and between monkey, ape, and human species.

Reconstruction of Ancient Environments

Reconstructing ancient environments is critical in understanding the organism's ecology, behavior, diet, and speciation events. The environment is never static. As noted earlier in the chapter, the earth's climate has changed many times in its history with dramatic effects. There are techniques used to obtain a chronology of these climate changes and, simultaneously, we get a better knowledge of how these changes affected ecology, behavior, diet, and speciation events. Two of these techniques are (1) stable oxygen isotope studies and (2) stable carbon isotope studies and animal remains.

STABLE OXYGEN ISOTOPE STUDIES

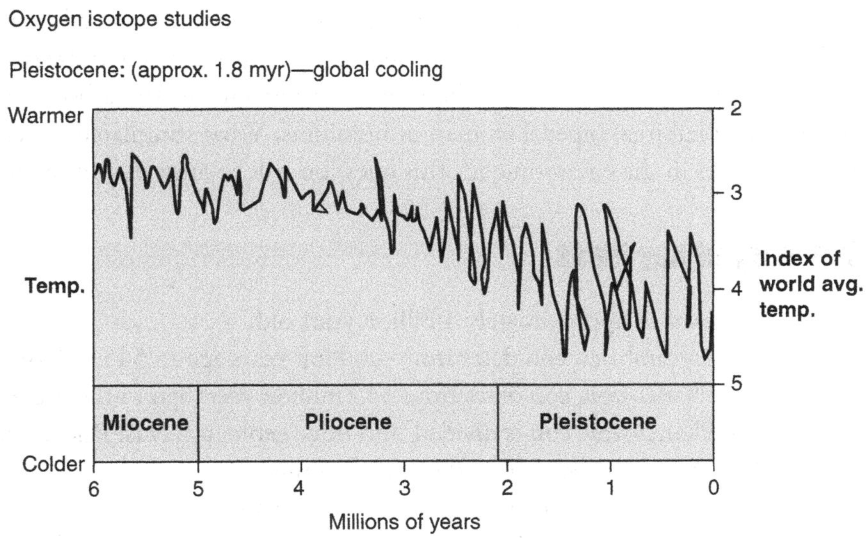

From the **Miocene** (25 to 6 million years ago) to the **Pleistocene** (2 million to 10,000 years ago), global temperatures have fluctuated from warmer to cooler to warmer (see the accompanying chart). This chronology was formulated based on the data from **stable oxygen isotope** studies. Most students know that atmospheric temperature affects ocean temperature. What they may not know is that changes in ocean temperature affect the amount of the oxygen-18 isotope in ocean water. Specifically, when the ocean temperature declines, the amount of oxygen-18 increases, and when the temperature increases, the amount of oxygen-18 decreases. This increase and decrease of the oxygen-18 isotope is recorded in the shells of ocean-dwelling microorganisms known as **foraminifera**. In order to obtain these shells for analysis, paleoclimatologists and geologists must extract sediment from the ocean floor.

STABLE CARBON ISOTOPE STUDIES AND ANIMAL REMAINS

When using animal remains in stable isotope studies, you are indirectly obtaining climate information by finding out the diet (trees, bushes, shrubs, or grass) of herbivores, since certain plant species are associated with specific environments. Edible plants employ either C3 or C4 photosynthesis. In other words, these types of photosynthesis determine how the plant extracts and uses carbon from atmospheric carbon dioxide (Ungar and Sponheimer 2011; Larson 2010).

When animals eat plants, **stable carbon-12** and **carbon-13 isotopes** are absorbed into their bones. In order to determine the prehistoric herbivore's diet and, indirectly, the climate in which it lived, paleontologists determine the ratio of carbon-12 to carbon-13 from the bones. A high ratio of carbon-12 to carbon-13 (C_4 plants include

(a)

(b)

Figure 7.4a,b. Tropical grassland savanna (a) and woodland forest environments (b).

savanna grasslands) is found in dry, tropical savanna grasslands, and a lower ratio (C_3 plants include trees, bushes, and shrubs) is found in wet, woodland environments associated with trees, bushes, and shrubs (Figure 7.4a,b).

As the Miocene came to a close, the environment in Africa became cooler. At the same time, we see the emergence of habitual bipedal animals or hominins. What stimulated the emergence of this new type of creature? Was it changes in the environment? This question will be answered in the next chapter.

Summary of Key Ideas

1. The earth is approximately 4 billion years old.
2. The Precambrian eon dates from 4 billion years ago to 545 million years ago.
3. The Phanerozoic eon dates from 545 million years ago to the present.
4. The Phanerozoic eon is divided into three geological eras: Paleozoic, Mesozoic, and Cenozoic.
5. The Mesozoic is known as the age of reptiles.
6. The Cenozoic is known as the age of mammals.
7. Three are two categories of dating methods: relative dating and numerical dating.
8. Relative dating does not give you a number; it can only tell you if the artifact is older or younger.
9. Numerical dating gives you a number; some dates are based on radiometric decay.
10. Half-life is the time it takes for half of an element to decay into a more stable form.
11. Reconstruction of ancient environments enables scientists to accurately explain prehistoric organisms' ecology, behavior, and diet, and speciation events.

Key Terms

Amino acid dating
Biostratigraphy
Carbon-14
Cenozoic
Chemical dating
Foraminifera
Genetic dating
Half-life
Isotope
Law of superposition
Mesozoic
Miocene
Molecular clock
Numerical dating
Paleozoic
Pangaea
Phanerozoic eon
Phylogenies
Pleistocene
Potassium-40/argon-40
Precambrian eon
Radiometric
Relative dating
Stable carbon-12 isotope
Stable carbon-13 isotope
Stable oxygen isotope
Strata
Stratigraphy
Thermoluminescence
Volcanic rock

Further Readings

Lee-Thorp, J. A., M. Sponheimer, and N. J. van der Merwe. 2003. What do stable isotopes tell us about hominid dietary and ecological niches in the Pliocene? *International Journal of Osteoarchaeology* 13: 104–13.

Lewin, R. 1997. *Bones of contention: Controversies in the search for human rigins,* 2nd ed. Chicago: University of Chicago Press.

Taylor, R. E. 1995. Radiocarbon dating: The continuing revolution. *Evolutionary Anthropology* 4: 169–81.

Chapter 7: Dating Methods

Study guide: By knowing the material below you will be well prepared for the exam. Don't forget to read the chapter and study your notes (this includes the chapter worksheets).

1. Be able to name the two categories of dating methods and give one example of each
2. Be able to explain stratigraphy, biostratigraphy, and chemical dating
3. Be able to explain Carbon 14 dating
4. Be able to list and explain the three (3) numerical (non-radiometric) dating techniques
5. The Earth is 4.6 billion yrs. old? How old are modern humans?
6. What was the dominant form of life during the Mesozoic?
7. What did the extinction of dinosaurs do for mammals?
8. Cenozoic Era is commonly known as__________.
9. Define relative dating. Be able to name and explain two (2) relative dating techniques
10. Define numerical dating. Be able to name and explain two (2) numerical dating techniques
11. What is biostratigraphy?
12. What is an isotope?
13. Give an example of a *primitive* mammalian trait retained by primates
14. The Phanerozoic eon is divided into three geological eras. List the eras with their respective time periods. In which era were all the continents joined together? What is the name for this landmass?
15. What are foraminifera? How are they related to stable oxygen isotope studies?
16. How are stable Carbon-12 and Carbon-13 isotopes used to determine prehistoric herbivore diet and climate (indirectly)? Discuss.

NAME ______________________________ DATE ____________

Chapter 7: Dating Methods

HOME WORK EXERCISES (10PTS.)

The goal of the home work exercises is assess your knowledge of the concepts and ideas presented in the chapter. Please answer the questions in each section as you go. When class lecture on this chapter is completed, carefully tear the home work exercises from the textbook and turn it in. **[Bonus pts. = 5]**

Use a pen or pencil to answer the questions below. Make sure your writing is LEGIBLE.

Section 7.1: Geological Time Period

1. The Phanerozoic eon is divided into three geological eras. List the eras with their respective time periods. In which era were all the continents joined together? What is the name for this landmass?

2. How might living organisms' evolution have been affected by landmasses' movements over time?

3. In which geological era did mammals diversified?

Section 7.2: Dating Methods

1. Differentiate between relative dating and numerical dating (radiometric) techniques, giving one example of each.

2. Define isotope. Is C-14 a stable or unstable isotope? Discuss how C-14 dating works.

3. Differentiate between numerical dating (radiometric) and numerical dating (nonradiometric) techniques, giving one example of numerical dating (nonradiometric).

Section 7.3: Reconstruction of Ancient Environments

1. What are foraminifera? How are they related to stable oxygen isotope studies?

2. How are stable Carbon-12 and Carbon-13 isotopes used to determine prehistoric herbivore diet and climate (indirectly)? Discuss.

CHAPTER 8

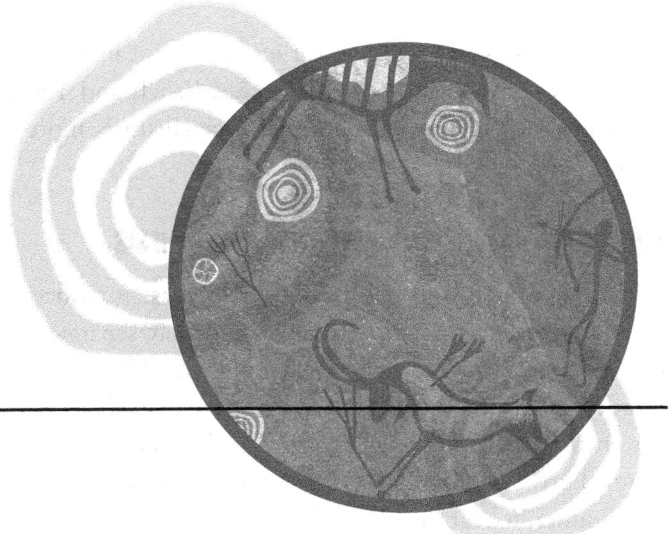

Australopiths and Early *Homo*

The Origin of Bipedalism

The genus *Australopithecus* is recognized as early, habitual bipedal human ancestors, although they cannot be considered *human*. The name *Australopithecus* was assigned to a fossil skull with small canines and a forward positioned foramen magnum in 1924 by Raymond Dart, professor of anatomy at the University of Witwatersrand in Johannesburg, South Africa (Dart 1925). While the important behavior of habitual bipedalism elevates **australopithecines** to the status of **hominins** (bipedal human ancestors), it takes much more than habitual bipedalism to be considered human. For instance, to be human is to depend on culture. We depend on our automobiles, telephones, and computers, to name a few. Some early hominins, within the genus *Homo*, for example, depended on stone tools, animal skins for clothing, and shelters. There is no strong evidence to date of stone tools and shelters in the australopith archaeology. In addition, to be human is to hunt cooperatively with weapons and, later, depend on agriculture. Again, substantive evidence of these complex behaviors are absent in the australopith archaeology. Nonetheless, the australopith archaeology does give us some clues to the roots of humanity.

What were the behavioral and environmental factors that led to **habitual bipedalism**? This is a very difficult question to answer simply because there is not enough primate fossil material from the late Miocene (7 to 6 million years ago) when the transition to habitual bipedalism would have occurred. Over the years, several hypotheses have been proposed for the origins of bipedalism. Let us now look at some of these hypotheses.

CHARLES DARWIN'S TOOL USE AND HUNTING HYPOTHESIS FOR BIPEDALISM

Drawing from the fact that Linnaeus placed humans in the order Primates because of the remarkable similarity between humans and apes, Darwin speculated that the shift to bipedalism was fairly easy. He reasoned that the transition from life in the trees to the ground involved changes in the foot, such as a nondivergent toe and longitudinal arch. Equally important, the hands and arms were free to manufacture tools and weapons. Overall, Darwin noted, "The free use of the arms and hands was partly the cause and partly the result of man's erect posture" (Darwin 1871/1981, p. 144).

ENERGY EFFICIENCY AND BIPEDALISM HYPOTHESIS

A second hypothesis in the arguments on the origin of bipedalism states that it evolved because it was more energy efficient in a woodland/savanna environment. As global temperatures cooled during the late Miocene-**Pliocene**

(7 to 5 million years ago), the earliest hominins had to evolve energy-efficient locomotor abilities to move rapidly between widely separated food patches (Isbell and Young 1996; Rodman and McHenry 1980). Compared to the average quadrupedal mammal, bipedalism is more energy-efficient walking at the same speed (Campbell et al. 2006).

BODY TEMPERATURE AND BIPEDALISM HYPOTHESIS

In this hypothesis, bipedalism evolved because it reduced heat buildup and facilitated heat dissipation in the earliest hominins as they moved between widely separated food patches in the middle of the day in order to avoid predators (predators hunt at dawn and dusk). Overexposure to solar radiation leads to heat stroke. Keeping this in mind, a quadruped—in the middle of the day when the sun is directly overhead—has more surface area exposed (its entire back) to the sun than a biped (Wheeler 1993). A biped has minimal surface exposure (shoulders and head). Furthermore, heat is trapped under the quadruped's body while the cooler air above flows over the biped's skin, aiding in heat loss through convection.

MALE PROVISIONING HYPOTHESIS FOR BIPEDALISM

This is a complex hypothesis. The author responds to the question of the origin of bipedalism inductively: monogamy, reduced birth spacing, and increase in population size were all the result of bipedalism (Lovejoy 1981). If males provided food for females, argues Owen Lovejoy (1981), females would not have to go out and forage for food. This reduces the chance of them getting eaten by predators or pregnant females suffering miscarriages due to stress caused by overexertion. In addition, this frequent provisioning of females would allow lactation and ovulation to occur simultaneously (as in modern human females), reducing birth spacing (in nonhuman primate females when there is lactation, there is no ovulation). In short, males' provisioning of females and offspring suggests the beginnings of monogamy, and the reduction in birth spacing results in more births and increased population size. According to Lovejoy (1981), male provisioning and the subsequent important behaviors it led to could not have occurred without bipedalism.

Pre-Australopithecines

Reconstruction of the climate between 8 to 7 million years ago shows a general cooling in Africa and the emergence of animals with both apelike and hominid-like traits (based on analysis of the long bones and pelvis). Interestingly, Ardipithecus—more than Sahelanthropus and Orrorin—"exhibits many important derived characters that are shared exclusively only with later hominids" (White et al. 2009, p. 84). These fossils are *Sahelanthropus tchadensis*, *Orrorin tugenesis*, and *Ardipithecus ramidus*. The critical question is, which one is within the lineage leading to the early hominin genus Australopithecus? Or, which one is within the earliest lineage leading to humans?

SAHELANTHROPUS TCHADENSIS (APPROXIMATELY 7 TO 6 MILLION YEARS AGO)

Sahelanthropus tchadensis was discovered in Chad in Central Africa. It is unusual because (1) it comes from the middle of Africa, approximately 2,000 miles west of where all early hominins in East Africa had been found in the past century; (2) it is surprising old: from 7–6 million years ago; and (3) it has a unique mix of ape and hominin traits, that is, the foramen magnum or hole that transmits the spinal cord to the base of the brain is positioned forward, suggesting bipedalism, but the cranial capacity (approximately 350 cc) is within the ape range.

ORRORIN TUGENESIS (APPROXIMATELY 6 MILLION YEARS AGO)

Orrorin tugenesis was discovered in Kenya in East Africa. It also has both ape and hominin traits. For instance, the dental enamel is thick and the thigh bone (femur) is angled inward. Both of these characteristics are found in hominins. In contrast, the arm and finger bones suggest adaptation for climbing. Reconstruction of the environment where the fossil bones were found (or in which the individual lived 6 million years ago) suggests a woodland/savanna environment.

ARDIPITHECUS RAMIDUS (APPROXIMATELY 5.8 TO 4.4 MILLION YEARS AGO)

Ardipithecus ramidus shares primitive characters with earlier hominin and apes (Figure 8.1 and Figure 8.2). As such, this fact reinforced the popular belief that the last common ancestor of chimpanzees and humans must have been a knuckle walker and vertical climber like chimpanzees and gorillas. But after 15 years of thorough analysis of the skeletal remains of Ardipithecus, this interpretation has been challenged (White et al. 2009). Specifically, the hand, foot, pelvic, vertebral, and limb structures and proportions were different compared to extant apes. Consequently, the last common ancestor was not apelike in locomotor or social behavior.

Cranial and Dental Anatomy (see Figures 8.2 and 8.3)

1. Cranial capacity (300cc to 350cc)
2. Small brow ridge
3. Midfacial projection
4. Lower face projection (but not to the extent of chimpanzees)
5. Small, nonprojecting canines (suggesting no male-male competition for females)
6. **Nonhoning complex**
7. Small molars (suggesting a diet of woodland resources—shrubs, fruits, etc)
8. Intermediate enamel thickness—between humans and chimpanzees (suggesting a non-abrasive diet but may have included hard foods) [Figure 8.3]

Postcranial Anatomy (see Figures 8.1 and 8.4)

Hand

1. Short metacarpals
2. No knuckle walking grooves
3. Large robust thumb (small in apes)
4. Long phalanges (fingers)
5. Flexible wrist (capable of flexing the hand up and down, which is impossible in apes)

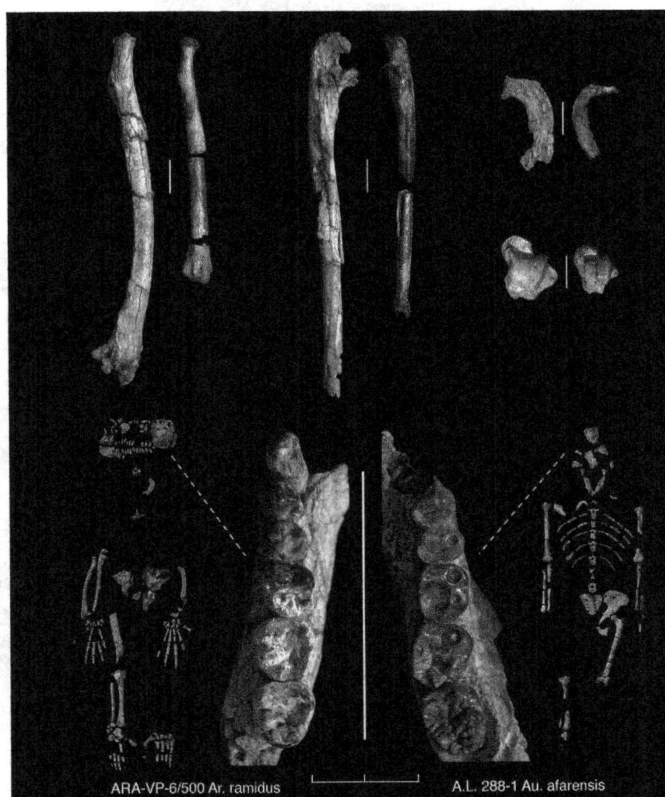

Figure 8.1. Comparisons of the cranial and postcranial of *Ardipithecus* (left) and early *Australopithecine* (right)

Original fossils in National Museum of Ethiopia, Addis Ababa.
Photo Credit: © 2009 Tim D. White

Humerus/Ulna/Radius

1. Lacks any characters diagnostic of habitual suspension (hanging/swinging) or vertical climbing
2. Upper and lower limbs same size (like monkeys)

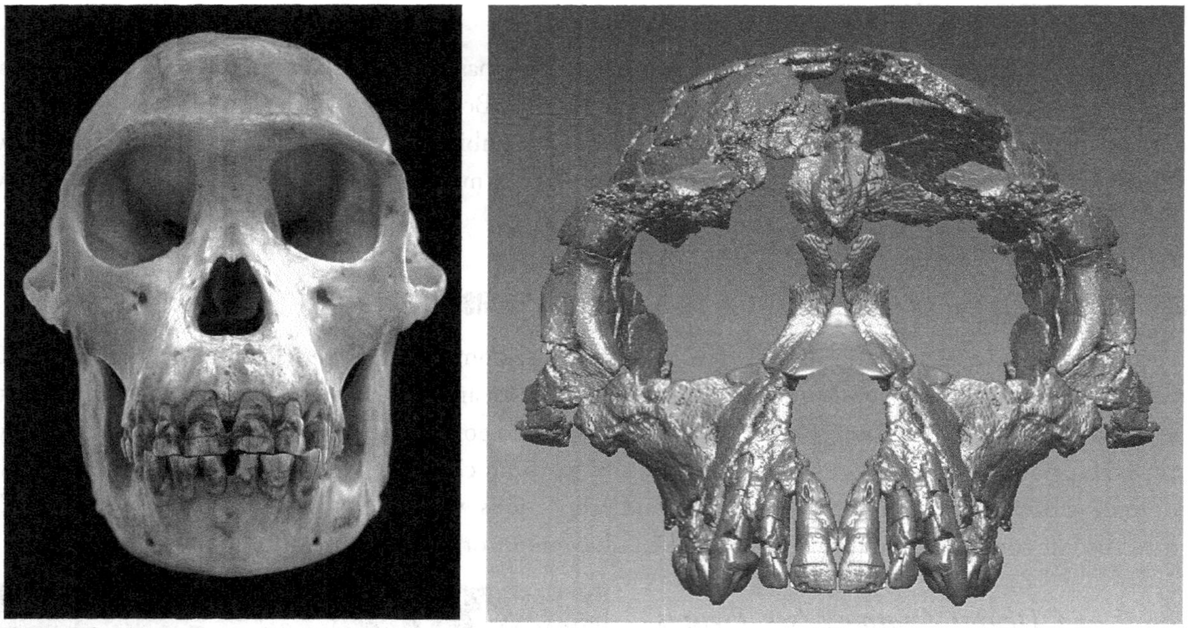

Figure 8.2. The skull of *Ardipithecus ramidus* (right) compared to the skull of a female chimpanzee (left)

Artifact Credit: Division of Mammals, National Museum of Natural History (NMNH), Smithsonian Institution. © 1996 David L. Brill
Fossil Credit: Original fossils in National Museum of Ethiopia, Addis Ababa. Digital Render Credit: © Gen Suwa, University of Tokyo

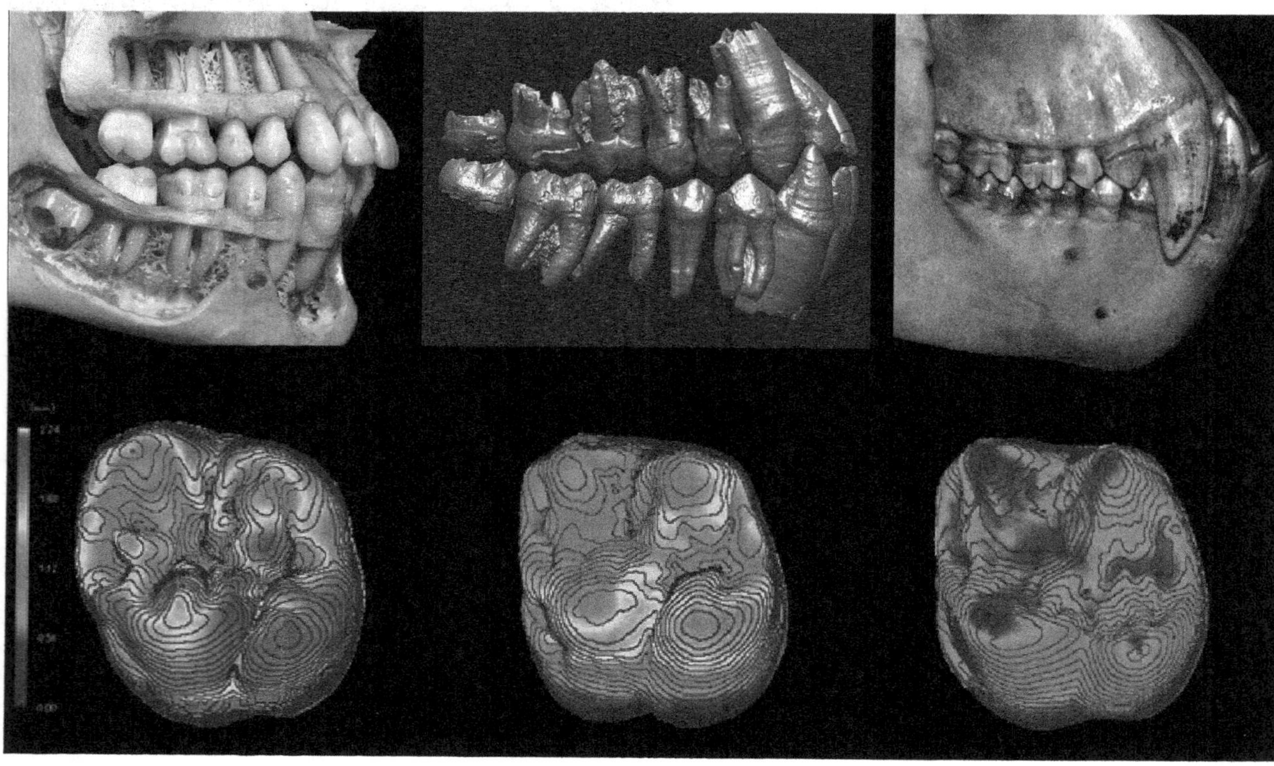

Figure 8.3. *Ardipithecus ramidus* dental views (center) compared with human dental views (left) and chimpanzee dental views (right)

Original fossils in National Museum of Ethiopia, Addis Ababa. Photo & Digital Render Credit: © by Tim D. White & Gen Suwa

Pelvis

1. Short and broad (suggesting bipedalism)
2. Anterior-inferior iliac spine present (not found in apes, but found in humans)
3. Slight sciatic notch present
4. Ischial tuberosity not angled upward (suggesting retention of climbing ability like apes)

Foot

1. Divergent big toe
2. Second through fifth metatarsals and phalanges are not curved
3. Robust lateral peroneal complex (suggesting substantial plantarflexion—walking on foot)

Overall, *Ardipithecus ramidus* was a biped on the ground and a **palmigrade** quadruped climber in the trees that lacked specializations for vertical climbing and knuckle walking. Furthermore, this species had small brains; small canines, and small molars. More importantly, the Ardipithecus skeleton indicates that chimpanzees cannot be used as a model for the earliest hominin social or locomotor behavior.

Reconstruction of the environment where the fossil bones were found (or in which the individual lived 5.8 million years ago) suggests a woodland environment.

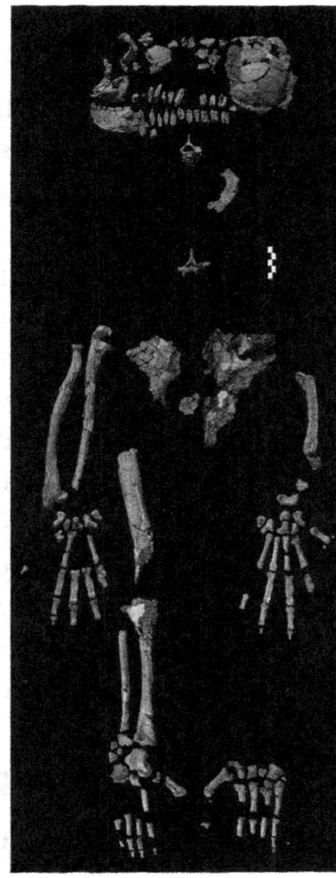

Figure 8.4. The *Ardipithecus ramidus* skeleton

Fossil Credit: Housed in National Museum of Ethiopia, Addis Ababa. Photo Credit: ©Tim D. White 2009.

Australopithecines

Australopithecine fossils, to date, have only been found in East Africa and South Africa. The East African fossils are well dated compared to the South African fossils. The geology of South Africa is comprised of limestone, as opposed to volcanic rock in East Africa, which makes radiometric dating difficult.

EAST AFRICAN AUSTRALOPITHECINES

Australopithecus anamensis *(Approximately 4 Million Years Ago)*

Australopithecus anamensis is the oldest australopithecine from Kenya in East Africa. Despite the fact that *anamensis,* like the pre-australopithecine fossils, has both hominin-like and apelike traits, it is still assigned to the genus *Australopithecus.* The fossil evidence consists of teeth, jaw, and skull fragments. Hominin traits include thick dental enamel, and the leg bone fragments suggest habitual bipedalism. In contrast, the skull fragments, large canines, and parallel tooth-row in the upper jaw are similar to those of modern apes.

Australopithecus afarensis *(Approximately 3.6 to 3.0 Million Years Ago)*

Australopithecus afarensis is the most celebrated of the primitive hominin species. The fossil is a 40 percent complete female skeleton, nicknamed "**Lucy**," discovered in Ethiopia in East Africa (Figure 8.5). The fossil evidence consists of cranial fragments, teeth, and postcranial (below the head) bones:

Skull (approximately 430 cc):

a. Projecting face
b. Receding forehead
c. Large brow ridges
d. Small canines
e. Parabolic tooth row

Hand:

a. Curved hand phalanges (may have still used the trees despite habitual bipedalism)

Humerus/ulna/radius (upper and lower arm bones):

a. Apelike ratio of arm and leg length (arms longer than legs)

Pelvis:

b. Partial bowl-shaped pelvis (adaptation for habitual bipedalism)

Femur/tibia (thigh/shin bones):

c. The femur shows a bicondylar angle (slanting inward at the knee joint), which is consistent with habitual bipedalism.
d. The proximal end of the tibia has a large shelf to absorb the stress of body weight during bipedal walking.

No foot bones were recovered for *Australopithecus afarensis*. Indirectly, however, we can say that *afarensis* had a nondivergent big toe and a well-developed arch, based on fossil footprints dated around the same time as *afarensis* that were found at the Laetoli site in Tanzania (Figure 8.6).

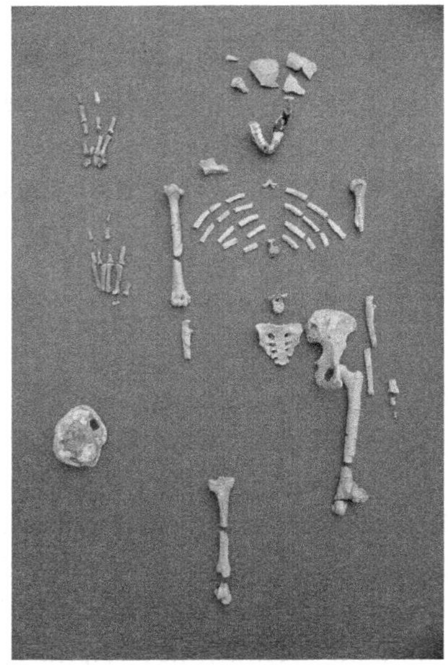

Figure 8.5. Australopithecus afarensis or 'Lucy."

© Alain Nogues/Sygma/Corbis

SOUTH AFRICAN AUSTRALOPITHECINES

Australopithecus africanus *(Approximately 3.0 Million Years Ago)*

Australopithecus africanus fossils were discovered in many regions in South Africa. They are very similar to *afarensis* (see traits above).

AUSTRALOPITHECUS SEDIBA (APPROXIMATELY 2.0 MILLION YEARS AGO)

In 2008, paleoanthropologist Lee Berger of the University of Witwatersrand in Johannesburg, South Africa and his team discovered two partial skeletal elements (later assessed as juvenile male and adult female) in Malapa Cave in South Africa. The fossils are commonly designated as Malapa hominids (MH) 1 and 2. On subsequent analysis, the skeletal individuals were placed in the genus *Australopithecus* and the species *sediba* because the researchers found several anomalies: the endocast (an impression of the inner surface of the skull) of the juvenile male's brain is different than South African australopithecines; while the partial pelvises indicated bipedalism, the angle of the pelvic blades in the male and female are different compare to other australopithecines; and the heel bones are unlike any other early hominin indicating a unique form of bipedalism (Gibbons 2011; Carlson et al. 2011; Kibii et al. 2011; Zipfel et al. 2011).

Figure 8.6. Laetoli footprints: indirect evidence for bipedalism 3.6 million years ago.

John Reader/Photo Researchers, Inc.

Cranial Anatomy (see Figure 8.7)

1. Endocast: inner surface of partial cranium-(impressions left by the boy's brain)
   a. Small brain: 450cc
   b. Australopith-like convolutional patterns on the frontal lobes but also some foreshadowing of features of human frontal lobes
   c. Posterior repositioning of the olfactory bulbs
   d. Orbital frontal shape and organization cluster closer to human endocasts

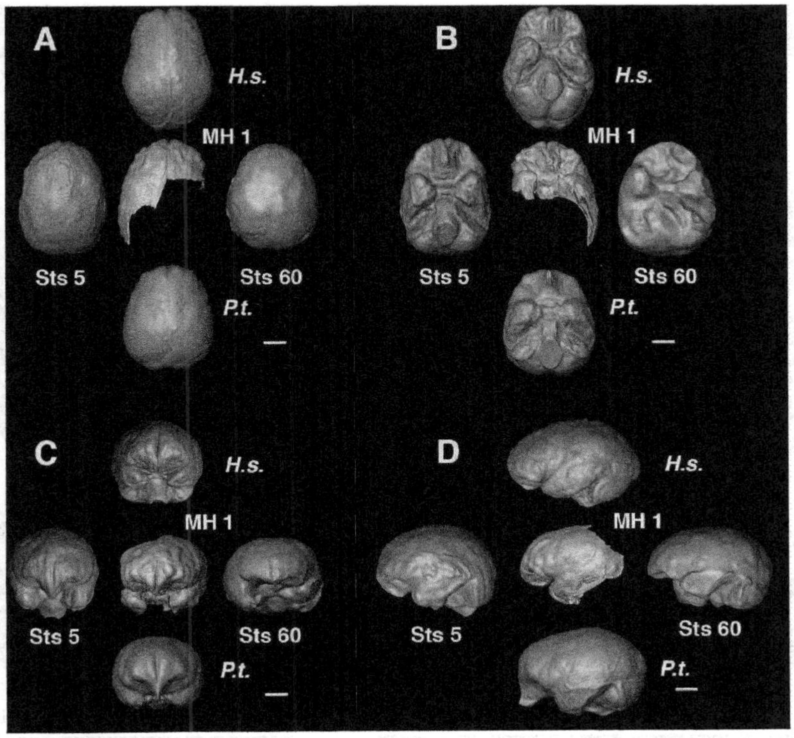

Figure 8.7. Comparisons of virtual endocasts

From *Science*, Volume 333, No. 6048, September 9, 2011 by Carlson, Stout, Jashashvili, de Ruiter, Tafforeau, Carlson, and Berger. Photo courtesy of Kristian J. Carlson.

Postcranial Anatomy **(see Figure 8.8a, b, c)**

1. Arm and hand (female adult)
   a. Long primitive arm
   b. Strong flexor apparatus associated with arboreal locomotion
   c. Long thumb and short fingers (like genus-*Homo*) associated with precision grip and tool production
2. Pelvis (partial pelvises of juvenile male and adult female)
   a. Large biacetabular diameter
   b. Small sacral and coxal joints
   c. Long pubic rami
   d. Vertical oriented and S-shaped iliac blades (like genus *Homo*)
   e. Greater robusticity of the iliac body (like genus *Homo*)
   f. Shortened ischia (like genus *Homo*)
   g. More superior oriented pubic rami (like genus *Homo*)

The researchers emphasize that these features in small-brained early hominins suggest that birthing of larger-brained babies was not driving the evolution of the pelvis in the Pliocene (Kibii et al. 2011).

3. Foot and Ankle (adult female)
   a. Ankle joint is mostly humanlike in form (some evidence for a humanlike arch and Achilles tendon)—more mobility
   b. More gracile calcaneal (heel) body (apelike)—small angled surface (rather than the broad flat heel bone in humans)
   c. More robust medial malleolus (apelike)

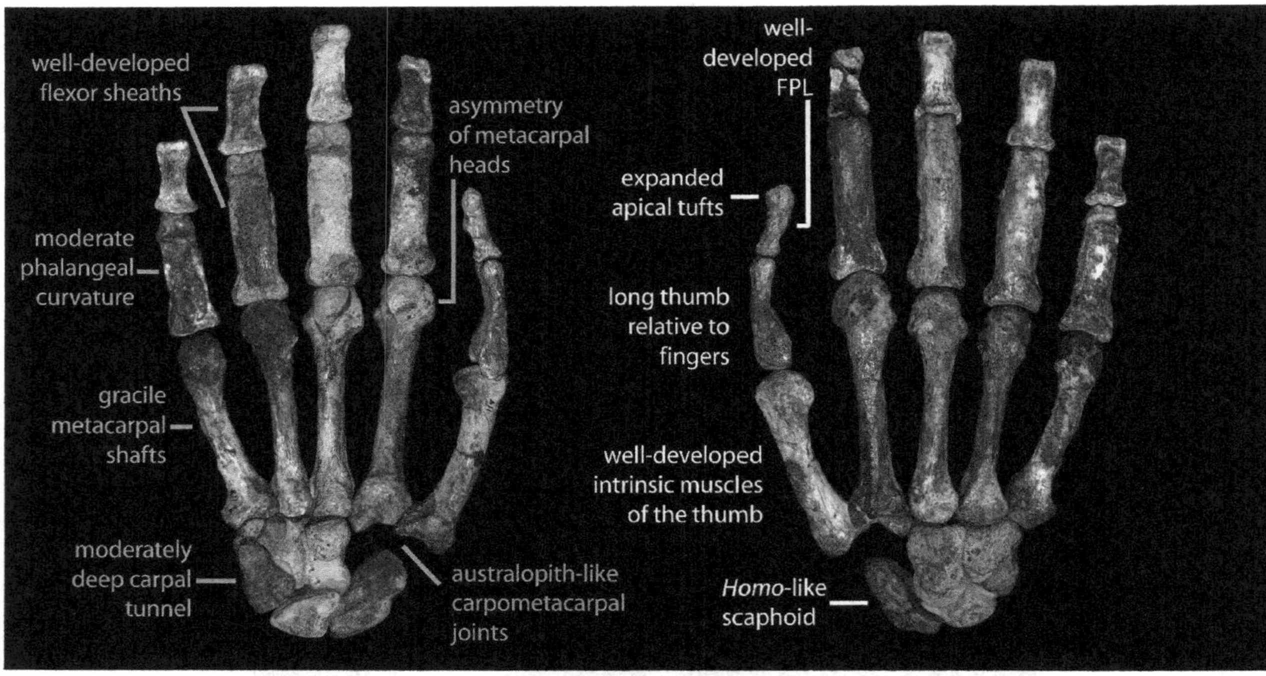

(a) From *Science*, Volume 333, No. 6048, September 9, 2011 by Kivell, Kibii, Churchill, Schmid, and Berger. Copyright © 2011 by the American Association for the Advancement of Science. Reprinted with permission from AAAS. Photo courtesy of Peter Schmid, University of Zurich, Wits University.

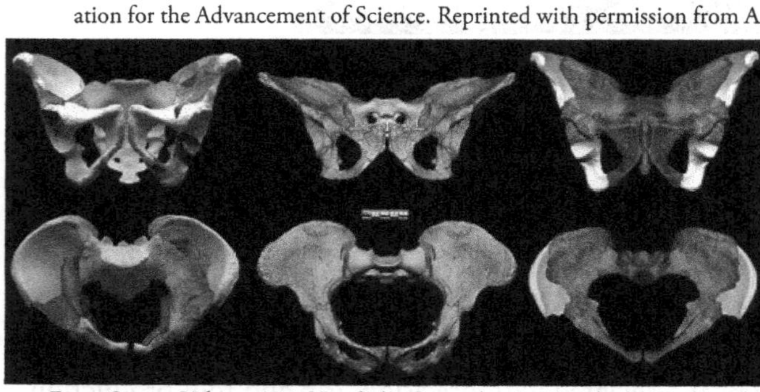

(b) From *Science*, Volume 333, No. 6048, September 9, 2011 by Kibii, Churchill, Schmid, Carlson, Reed, de Ruiter, and Berger. Copyright © 2011 by the American Association for the Advancement of Science. Reprinted with permission from AAAS. Photo courtesy of Peter Schmid, University of Zurich, Wits University.

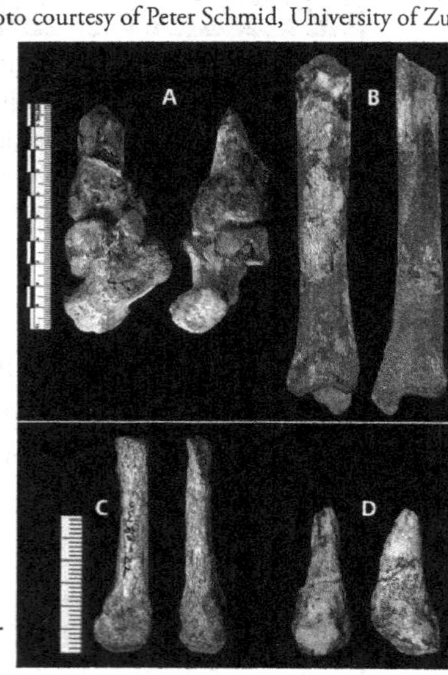

From *Science*, Volume 333, No. 6048, September 9, 2011 by Zipfel, DeSilva, Kidd, Carlson, Churchill, and Berger. Copyright © 2011 by the American Association for the Advancement of Science. Reprinted with permission from AAAS. Photo courtesy of Peter Schmid, University of Zurich, Wits University.

(c)

Figure 8.8a,b,c. (a) Australopithecus sebida hand, (b) partial pelvises of Australopithecus sebida, 1(left) and 2 (right) compared to another South African australopithecine (center), (c) foot and ankle of Australopithecus sebida

Robust Australopithecines (East and South Africa, Approximately 2.5 to 1.2 Million Years Ago)

According to the fossil record, the **robust australopithecines** appeared and disappeared fairly quickly. The three major species are *Australopithecus aethiopicus* from East Africa, *Australopithecus boisei* from East Africa, and *Australopithecus robustus* from South Africa (Figure 8.9a,b). The "robust" designation refers to their faces, jaws, and teeth (not body size):

a. Large molars, small incisors
b. Canines nonprojecting
c. Massive jaws/chewing surface
d. Wide projecting faces

The massive jaws and back teeth point to powerful chewing ability. This suggests that robust australopithecines relied regularly on harsh, vegetative foods. In essence, this herbivorous specialization might have led to their extinc-

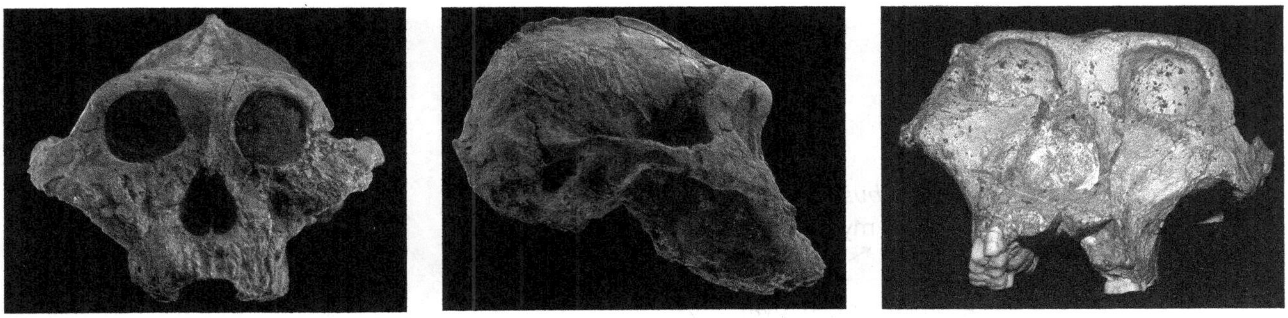

(a) (b)

Figure 8.9a,b. Australopithecus boisei—front and side views (a) and Australopithecus robustus (b).

(a) National Museum of Kenya, Nairobi. Photo Credit: © 1985 David L. Brill

(b) Transvaal Museum: Dept. Digital Information

tion when the environment changed and their favorite food disappeared. The evidence from dental anatomy shows that the less robust australopithecines had more flexibility in their diets because they subsisted on seeds, nuts, insects, and fruits. However, new research (Ungar and Sponheimer 2011) have challenged this traditional thinking of "robust" australopithecine diet. The researchers argue that tooth and jaw size, shape, and structure indicate what early hominins were capable of eating and suggest the selective pressures faced by their ancestors, rather than what specific individuals ate (see last section on Diets of Early Hominins).

Evidence of Tool Use in Australopiths?

AUSTRALOPITHECUS GARHI (2.5 MILLION YEARS AGO)

Australopithecus garhi was discovered in the Middle Awash region of Ethiopia in East Africa. It is a very controversial hominin because of the suggestion that this australopithecine species was the first tool maker and user. While primitive stone tools were found in the Middle Awash region, they were not associated with *Australopithecus garhi*, found in the same area. The *garhi* fossil assemblage consisted of cranial fragments, teeth, and some long bones:

Skull: approximately 450 cc
Dentition: second biscuspids have equal-size cusps
Humerus/femur (upper arm and thigh bones):
  a. Human-like ratio of arm and leg length (legs longer than arms)

If the primitive tools recovered in the Middle Awash can be linked, scientifically, to *Australopithecus garhi*, then the genus *Australopithecus* would be the makers and users of the first tools: the **Oldowan complex**. The greater implication is that australopithecines had *culture* and we can call them humans. Until remains of early *Homo* are found in this region, *Australopithecus garhi* was perhaps the first hominin to produce and use tools.

Australopithecus garhi at 2.5 million years ago and Australopithecus sediba at 2.0 million years ago: Implications for the Origin of Genus Homo

The ratio of arm (humerus) length to leg (femur) length; cranial, dental, and postcranial features; and its chronological position at 2.5 million years ago suggest that *Australopithecus garhi* was ancestral to *Homo* (Clark et al 1994; de Heinzelin et al 1999; de Heinzelin et al 2000; Larsen 2010). Tool production and use may have been passed down to *Homo*.

When *Australopithecus sediba* species is brought into the conversation, despite the human-like posterior repositioning of its olfactory bulbs; orbitofrontal shape similar to human endocasts; and human-like pelvic iliac blades they still could not themselves give rise to *Homo* (2.3 million years ago). They are too young. A phylogeny showing a hypothesis of evolutionary relationship is shown in figure 8.10.

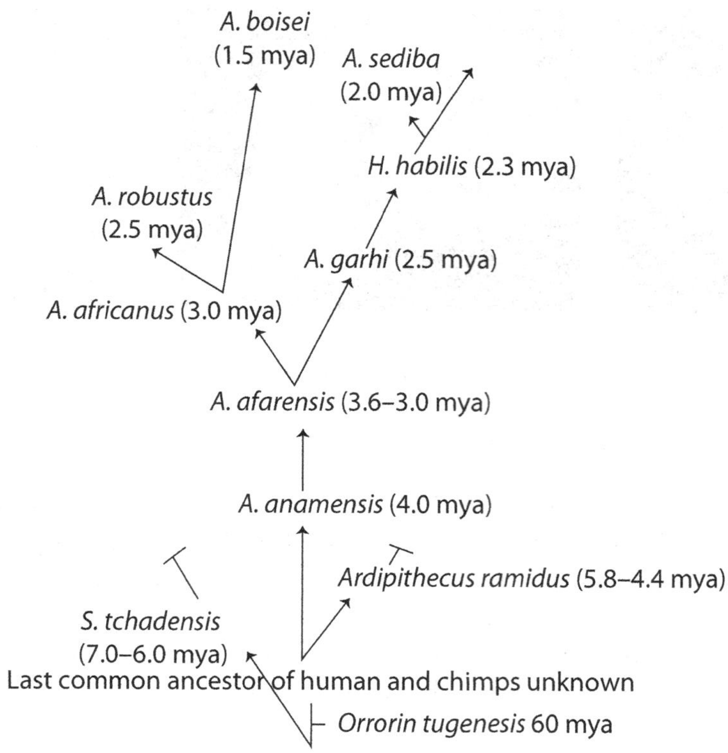

Figure 8.10. Hypothesis of hominin phylogeny

Early Homo

HOMO HABILIS (EAST AFRICA, APPROXIMATELY 2.3 TO 1.8 MILLION YEARS AGO)

Habilis fossils were discovered in Kenya and Ethiopia. The key difference between **Homo habilis** and the australopithecines is the larger cranial capacity of *Homo habilis* (Figure 8.11a,b). This is the reason why these fossils are in the genus *Homo*. This is where the path to humanness began:

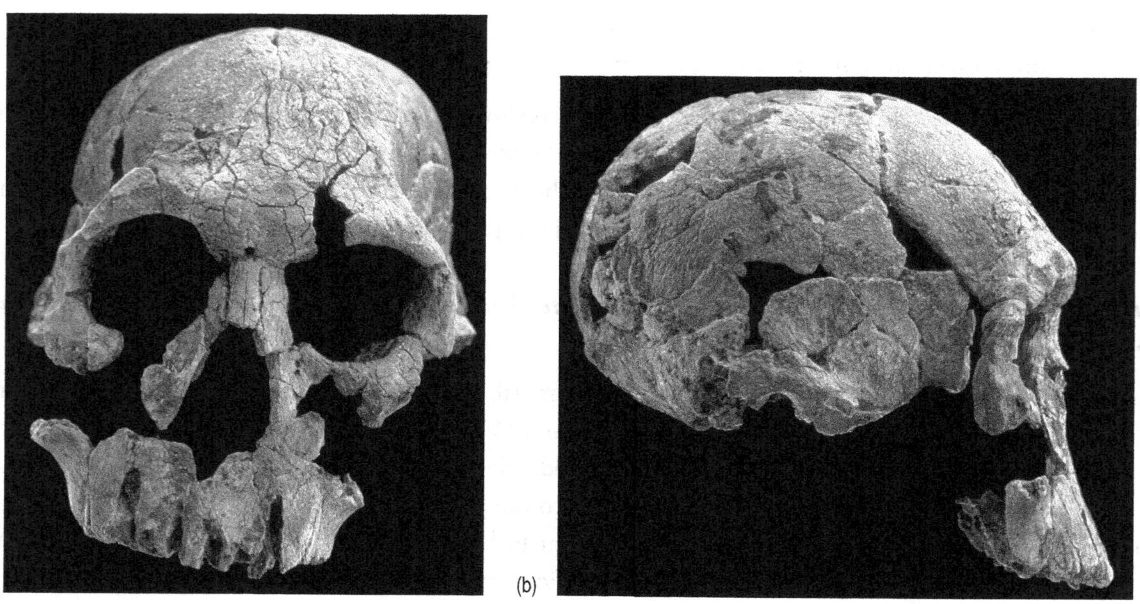

(a) (b)

Figure 8.11a,b. Homo habilis—front and side views.

National Museum of Kenya, Nairobi. Photo Credit: © 1985 David L. Brill

a. Skull: increased cranial capacity (approximately 650 cc on average; australopithecines approximately 450 cc on average)
b. Dentition: reduced size of face, teeth, jaws
c. Femur/tibia/foot: retained some climbing ability/shorter legs than arms
d. Social behavior: makers of the earliest stone tools—Oldowan complex (more likely than *Australopithecus garhi*) Scavengers, not hunters

Based on the larger cranial capacity, it is more likely that *Homo habilis* was the first maker and user of the first tools (rather than *Australopithecus garhi*). What can we say about the evidence from Middle Awash, Ethiopia? There might be a *Homo habilis* fossil in Middle Awash yet to be found. On the other hand, *garhi* could be a direct ancestor of *Homo habilis*. More excavations are needed to solve this riddle.

Diets of Early Hominins

In chapter 7, we discussed the technique of using animal remains in stable isotope studies in order to obtain information on prehistoric herbivore diet and climate. To review, stable carbon-12 and carbon-13 isotopes are absorbed into the bones of herbivores (and carnivores that eat herbivores). A high ratio of carbon 12 to carbon-13 is found in dry, tropical savanna grassland (C_4 plants) and a lower ratio is found in wet, woodland environments associated with trees, bushes, and shrubs (C_3 plants).

In a study by Peter Ungar and Matt Sponheimer (2011), stable carbon isotopes analysis of dental enamel and dental microwear analysis was done on early hominins to assess their diet. According to the researchers, dental enamel is preferred because it is highly mineralized and less susceptible to post-depositional alteration than bone. They challenge traditional belief among paleoanthropologists that tooth and jaw size, shape, and structure indicate what specific early hominin individuals ate.

Based on the most influential hypotheses in the paleoanthropology, thickly enameled, large, flat molars and heavily built crania and jaws of later 'robust' australopithecines reflect an adaptive shift from diets dominated by forest fruits to hard, brittle nuts or seeds or underground tubers. And, the earliest members of our own genus acquired a broad range of foods, including meat because they used tools. Early *Homo*'s face, teeth, and jaw were smaller compared to the australopiths. In contrast, Ungar and Sponheimer (2011) argue that tooth and jaw size, shape, and structure offer important clues to the fracture properties of foods and associated masticatory stresses and strains to which a species is adapted; they indicate what early hominins were capable of eating; and they indicate the selected pressures faced by their ancestors.

DENTAL MICROWEAR ANALYSIS

Mammals show a strong association between dental microwear pattern and food fracture properties (Ungar and Sponheimer 2011). Those mammals that crush hard, brittle foods, such as nuts and bones, for example, typically have occlusal microwear dominated by pits. Those that shear tough items, such as leaves and meat more often show long, parallel striations on their wear surface. What is interesting is that these pits and scratches are traces of actual chewing events reflecting diet in the days and weeks before death. Animal species with a more generalized diet should show a broader range of microwear textures than those that consume a more limited variety of foods. Table 8.1 shows a model for microwear formation based on data by Ungar and Sponheimer. Also, see figures 8.12 and 8.13.

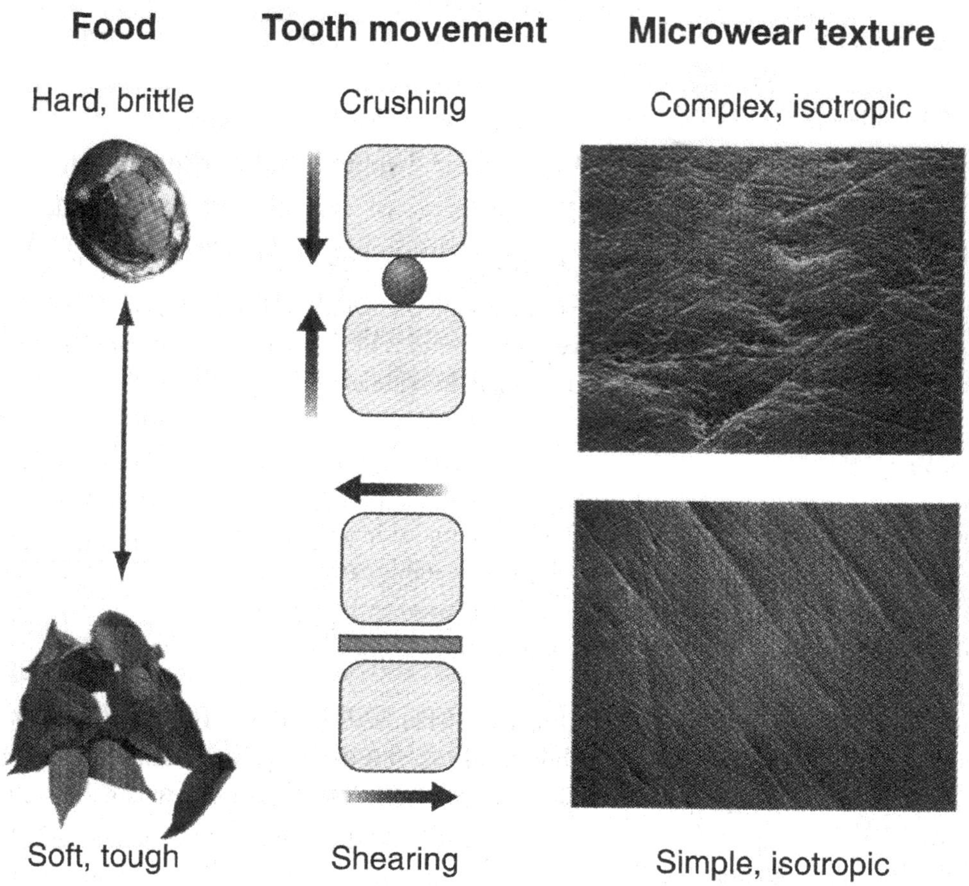

Figure 8.12. Microwear textures of early hominins

From *Science*, Volume 334, No. 6053, October 14, 2011 by Peter S. Ungar and Matt Sponheimer. Copyright © 2011 by the American Association for the Advancement of Science. Reprinted with permission from AAAS.

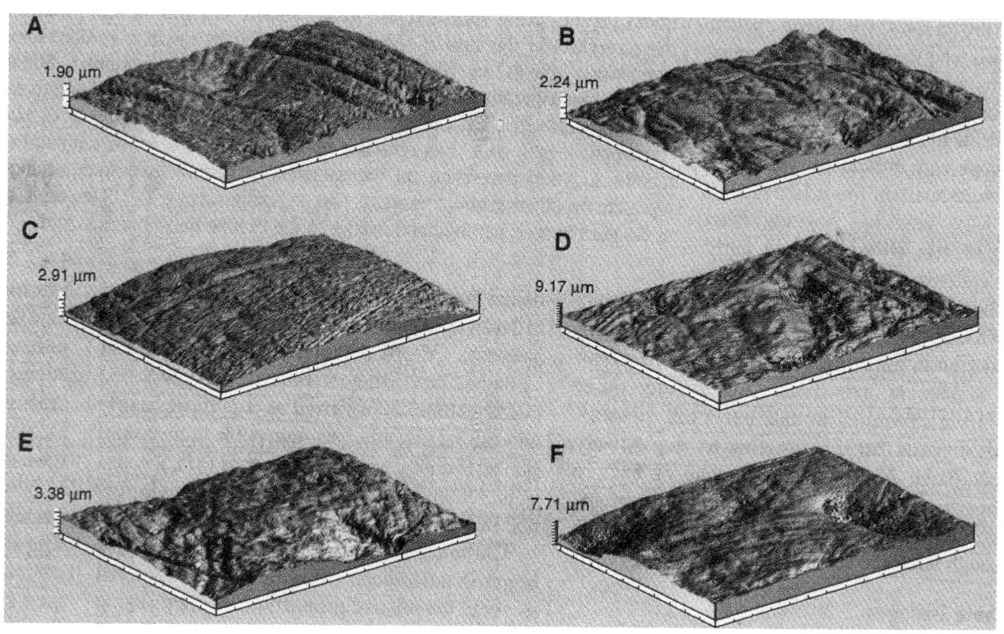

Figure 8.13. Photosimulations of microwear surfaces

From *Science*, Volume 334, No. 6053, October 14, 2011 by Peter S. Ungar and Matt Sponheimer. Copyright © 2011 by the American Association for the Advancement of Science. Reprinted with permission from AAAS.

Table 8.1: Model for microwear formation

Fossil species	**Food and tooth movement:** hard brittle foods crushed between opposing teeth causing pitting with complex surface textures	**Food and tooth movement:** soft and tough foods sheared between opposing teeth that slide past one another causing parallel scratches and simpler surface textures
afarensis	Low complexity	Low to moderate complexity (not as great as modern leaf eaters)
africanus	Low complexity	Low to moderate complexity (not as great as modern leaf eaters)
boisei	Low complexity	Low to moderate complexity (not as great as modern leaf eaters)
robustus	High complexity	Low complexity
H. habilis	Low complexity	Low to moderate complexity (not as great as modern leaf eaters)
H. erectus	High complexity	High complexity
anamensis	Low complexity	Low to moderate complexity (not as great as modern leaf eaters)

Source: Modified from Ungar and Sponheimer 2011.

STABLE CARBON ISOTOPES

Conventional wisdom states that early hominins' diets were very different compared to diets of living African apes or other herbivores. For example, chimpanzee diet is dominated by C_3 plants (trees, bushes, and shrubs) whether in preferred forest habitats or open savanna with abundant grasses. And, the diets of warthogs, hippos, and zebras, for example, are dominated by C_4 plants (tropical grasses). However, Ungar and Sponheimer research shows that the current view in the field is too simplistic. Data showing the stable carbon isotope results are in Table 8.2.

Table 8.2: Carbon isotope compositions

Fossil species	**Carbon isotope: C_3 plants** (trees, bushes, shrubs)	**Carbon isotope: C_4 plants** (tropical savanna grassland)
Ar. ramidus	100%	
africanus	50%	50%
Early *Homo*	50%	50%
robustus	50%	50%
boisei	15% to 25%	75% to 85% (similar to grass-eating warthogs, hippos, and zebras)

Source: Modified from Ungar and Sponheimer 2011.

Stable carbon isotope results indicate that the diet diversity within the hominins is complex. A good example is the so-called "robust" australopithecines having similar jaws and teeth but very different diets. This complexity might be caused by changing ecology due to global cooling, geography, and different time periods.

In conclusion, these microwear and stable carbon isotope studies have challenged the way we think about hominin diet, which is that early hominin craniodental functional morphology evolved for increasing consumption of hard, brittle foods as global temperatures cooled and savannas spread. The fact that *Australopithecus boisei* (East Africa) and *Australopithecus robustus* (South Africa) have different carbon isotope compositions, with the South African "robust" australopithecine consuming a much higher fraction of C_3 foods, challenges this thinking. (see Tables 8.1 and 8.2). In addition, we have some new insights into food choices and foraging strategies. For instance, we know that animals will 'fall back' on other foods when their preferred food is not available. Moreover, its morphological complex—whether specialized or not—may be able to serve more than one function. Ungar and Sponheimer (2011) argue that the "robust" morphology may have functioned in high-stress hand-object feeding for *A. robustus* but in repetitive loading grinding of tough foods for *A. boisei.*

Summary of Key Ideas

1. Australopithecines are early, habitual bipedal human ancestors.
2. All habitual bipedal primates are assigned to the category hominin.
3. There is no definitive evidence for culture or tool manufacture or use in the genus *Australopithecus.*
4. There are many behavioral and environmental factors that led to the origins of bipedalism.
5. The pre-australopithecines are important because they appear just before the split between human and chimpanzee lineages. One could be the "missing link."
6. Robusticity in some australopithecine species refers to face, jaws, and teeth, not body size.
7. The key difference between Homo habilis and the australopithecines is the larger cranial capacity of *Homo habilis.*
8. *Homo habilis* is the maker and user of the first tools.
9. The first tools are known as Oldowan tools (part of the Oldowan complex).
10. Ardipithecus ramidus was a bipedal on the ground and a palmigrade quadruped climber in the trees that lacked specializations for vertical climbing or knuckle walking.
11. Based on analysis of the Ardipithecus skeletal materials, apes cannot be used as a model for the earliest hominin social or locomotor behavior.

Key Terms

Ardipithecus ramidus
Australopithecines
Australopithecus afarensis
Australopithecus africanus
Australopithecus anamensis
Australopithecus garhi
Body temperature and bipedalism hypothesis
Darwin's tool use and hunting hypothesis
Energy efficiency and bipedalism hypothesis
Habitual bipedalism
Hominin
Homo habilis
Male provisioning hypothesis
Nohoning
Oldowan complex
Orrorin tugenesis
Palmigrade
Pliocene
Robust australopithecines
Sahelanthropus tchadensis

Further Readings

Conroy, G. C. 1997. *Reconstructing human origins.* New York: W. W. Norton.

Johanson, D., and E. Blake. 1996. *From Lucy to language.* New York: Simon & Schuster.

Lovejoy, C. O. 1988. Evolution of human walking. *Scientific American* 259: 118–25.

McHenry, H. M., and K. Coffing. 2000. *Australopithecus to Homo:* Transformations in body and mind. *Annual Review of Anthropology* 29: 125–46.

Ward, C., M. Leakey, and A. Walker. 1999. The new hominid species *Australopithecus anamensis. Evolutionary Anthropology* 7: 197–205.

White, T. D. 2002. Earliest hominids. In *The primate fossil record,* ed. W. C. Hartwig, 407–17. Cambridge, UK: Cambridge University Press.

Wolpoff, M. H. 1999. *Paleoanthropology,* 2nd ed. Boston: McGraw-Hill.

Wood, B. 1992. Origin and evolution of the genus *Homo. Nature* 355: 783–90.

Wood, B. 2002. Paleoanthropology: Hominid revelations from Chad. *Nature* 418: 133–35.

INTERNET RESOURCES

Toumai—The Human Ancestor
http://www.cnrs.fr/cw/fr/pres/compress/Toumai/Tounaigb/resumegb.html
This Web site was created by the scientists who found the *Sahelanthropus* fossil in Chad. It provides details about *Sahelanthropus.*

The Origin of Hominin Bipedalism
http://www.riverapes.com/AHAH/ComparativeBiology/Bipedalism/Bipedalism.htm
This Web site provides the current hypotheses for the origin of hominin bipedalism.

Origins of Humankind
http://www.pbs.org/wgbh/evolution/humans/humankind/a.html
This Web site is maintained by PBS. It gives basic information on several early hominin species.

Orrorin tugenesis
http://humanorigins.si.edu
This Web site provides anatomical information on *Orrorin tugenensis,* the 6 million-year-old fossil from Kenya.

Becoming Human
http://www.becominghuman.org
This Web site was created by the Institute of Human Origins. It provides information on the discovery of *Australopithecus afarensis* (Lucy).

Homo habilis
http://humanorigins.si.edu
This Web site provides illustrations of several *Homo habilis* specimens.

Chapter 8: Australopiths and Early *Homo*

Study guide: By knowing the material below you will be well prepared for the exam. Don't forget to read the chapter and study your notes (this includes the chapter worksheets).

1. Which came first: bipedalism or big brains?
2. When did the first hominids appear in Africa?
3. Know the body size of early hominids
4. Know the date of the oldest tools. What were they called? Who made them?
5. What is the scientific name (i.e., genus and species) for "Lucy?"
6. What region in East Africa is called the "cradle of humanity"?
7. Know the traits indicative of a biped
8. Be able to discuss the four (4) hypotheses for why hominids emerged. What are their strengths and weaknesses?
9. *Australopithecus africanus*—where was this fossil found?
10. Know the three (3) pre-australopithecine fossils and where they were found. Why are they important?
11. What is the Laetoli site in Tanzania famous for? What is significant about the artifact found there?
12. Know the differences between "robust" and "gracile" australopithecines. Based on new research are these interpretations totally correct? Discuss
13. What is a sagittal crest? What does it indicate?
14. Why is *Australopithecus garhi* important?
15. What is the date of *Australopithecus garhi*? How is this species related to early Homo?
16. In the analysis of *Ardipithecus ramidus*' postcranial anatomy, what is inferred about its locomotion and was it similar to the great apes?
17. What makes *Australopithecus sediba* different from other early hominins? What is the current controversy concerning *A. sediba*? Discuss.
18. What is the date of early *Homo*? Name one species of early *Homo*? Name one key trait that distinguishes early *Homo* from the australopithecines.

NAME ______________________________ **DATE** ______________

Chapter 8: Australopiths and Early *Homo*

HOME WORK EXERCISES

The goal of the home work exercises is assess your knowledge of the concepts and ideas presented in the chapter. Please answer the questions in each section as you go. When class lecture on this chapter is completed, carefully tear the home work exercises from the textbook and turn it in. **[Bonus pts. = 10]**

Section 8.1: The Origin of Bipedalism

1. Define hominin. What is one key trait necessary to be called a hominin?

2. What is name *Australopithecus*?

3. There are four hypotheses for the origin of bipedalism in the text. Using your knowledge of human origins at this point in the course, find weaknesses in each hypothesis.

Section 8.2: Pre-Australopithecines

1. Name the three (3) pre-australopithecines in human evolution. Why are they important in the origins of early hominins?

2. In the analysis of *Ardipithecus ramidus'* postcranial anatomy, what is inferred about its locomotion? Was its locomotion similar to the great apes?

Section 8.3: Australopithecines

1. Australopithecines are considered undisputed____________________________.

2. The scientific name for "Lucy" is____.

3. In "Lucy," name one cranial trait (above the head—this includes dental) and three post-cranial traits (below the head) that are similar to modern humans.

4. Australopithecines have been found in East and South Africa. *Australopithecus africanus* comes from____ _________________________.

5. What makes *Australopithecus sediba* different from other early hominins? What is the current controversy concerning *A. sediba*? Discuss.

6. Name the three (3) Robust australopithecines. Describe their cranial traits. What do these traits indicate? Based on new research are these interpretations totally correct? Why? [Note: see section on Diets of Early hominins]

Section 8.4: Evidence of Tool Use in Australopiths?

1. What is the date for *Australopithecus garhi*? In what country was the fossil found? Why is *Australopithecus garhi* important?

Section 8.5: Early *Homo*

1. What is the date of early *Homo*? Name one species of early *Homo*? Name one key trait that distinguishes early *Homo* from the australopithecines.

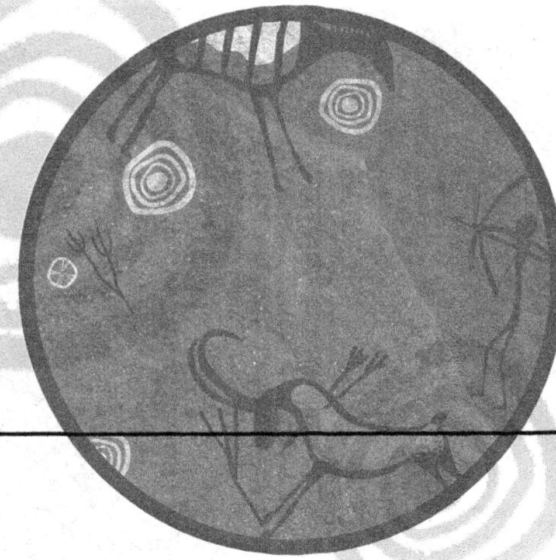

CHAPTER 9

Homo Erectus

Pithecanthropus erectus

In 1891, a human-like molar, a partial skull, and a complete femur were discovered in Java, Indonesia by a Dutch anatomist named **Eugene Dubois**. On analysis, he found that the skull was too large to belong to a modern ape and too small to belong to a modern human. Furthermore, the femur was very similar to a modern human's femur. Dubois concluded that the bones belonged to an upright-walking human ancestor. He called his fossil ***Pithecanthropus erectus*** (ape-man), a name created by Germany's preeminent anatomist, Ernst Haeckel (1834–1919). Later, the genus name *Homo* replaced *Pithecanthropus* because of the strong evidence of human-like behavior in this species based on its material culture, such as bifacial stone tools, modified animal bones indicating hunting and butchering, and burned flint suggesting the use of fire. It is proposed that these cultural innovations enabled ***Homo erectus*** to extend its geographical range and become the first hominin to leave Africa, based on fossils found in Europe and Asia. Details of *Homo erectus*'s morphology are presented in Figures 9.1 and 9.2.

Morphology of *Homo erectus*

Skull/face (Figure 9.1):

1. Cranial capacity (800 cc to 1,000 cc)
2. Large brow ridge
3. Sloping forehead
4. Projecting face (less projection and smaller than australopithecines)
5. No chin
6. Smaller teeth and jaw than australopithecines
7. Higher skull
8. Skull broader toward the bottom

Postcranial (figure 9.2):

Homo erectus was very large compared to australopithecines. One can speculate that high-quality nutrition such as meat was instrumental in their larger body size. Their body proportions were similar to modern humans. This evidence comes from one of the most complete hominin skeletons found at Nariokotome, on the west side of Lake Turkana in Kenya. This is an 80 percent complete juvenile skeleton with a 900 cc cranial capacity: (Figure 9.3).

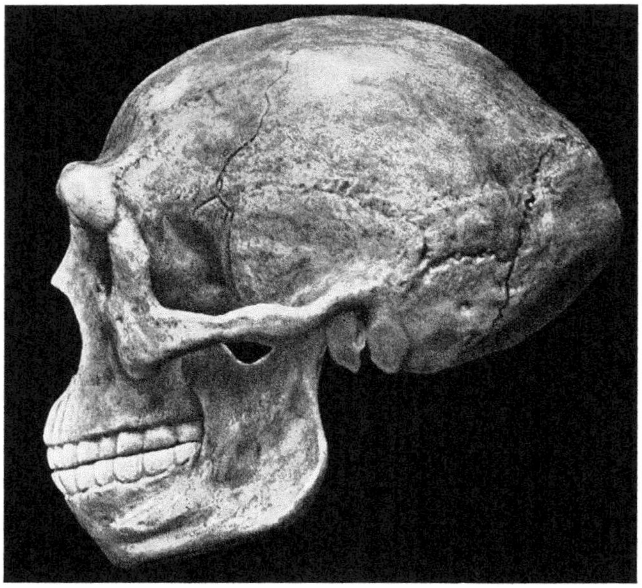

Figure 9.1. Homo erectus fossil from Zhoukoudian, China—side view.

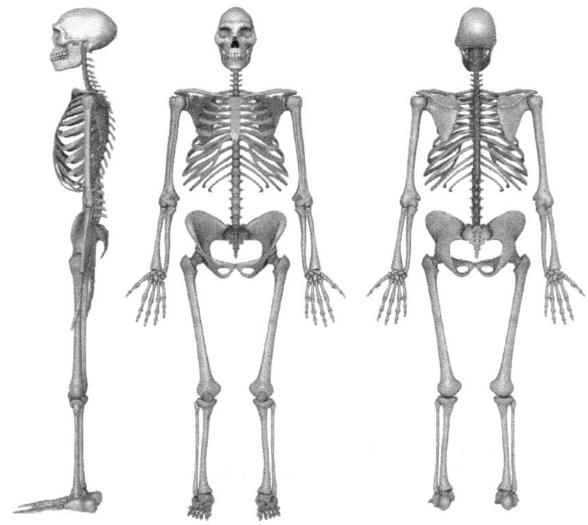

Figure 9.2. The postcranial of Homo erectus.

Nariokotome boy (1.6 million years old):

1. Long arms and legs/slender torso
2. Bodily proportions identical to tropical sub-Saharan Africans
3. Rib cage (thoracic vertebrae) slightly different compared to modern humans

Geography

Homo erectus in Africa, Asia, and Europe (Approximately 1.8 Million to 300,000 Years Ago)

Homo erectus fossil material is found as far as Asia by 1.8 million years ago. *Homo habilis* is not found in the African fossil record after 1.8 million years. One can speculate that *Homo habilis* evolved into *Homo erectus*. This fits the fossil evidence indicating that the earliest record of *Homo erectus* in Africa is approximately 1.8 million years ago. Three hundred thousand years (1.5 million years) later, the australopithecines were extinct, not found in the fossil record. It is possible that they lost in competition with the big-brained, larger, and technological advanced *Homo erectus* species. Very soon after their emergence, the *Homo erectus* species moved out into the Old World, reaching as far as Asia and Indonesia. Their designation as "the first world travelers" is based on evidence of *Homo erectus* fossils, fossils and tools, or tools found at the following sites: Ethiopia (Buia, Daka, Melka Kunturé, Middle Awash, Omo); Kenya (Koobi Fora, Lainyamok, Nariokotome, Olorgesailie, Turkana); Tanzania (Olduvai, Kalambo Falls); Morocco (Ternifine, Thomas Quarry, Salé); Israel (Gesher Benot Ya'agov, Ubeidiya); England (Hoxne); France (Abbeville, St. Acheul, Terra Amatta); Spain (Aridos, Gran Dolina, Torralba/Ambrona, Venta Micena); Italy (Ceprano); Republic of Georgia (Dmanisi); China (Gongwangling, Hexian, Lantian, Zhoukoudian); and Indonesia (Ngandong, Sambungmachan, Sangiran, Trinil) (Campbell et al. 2005; Day 1965; Larsen 2008; Relethford 2005). Migration to the colder regions would have been very difficult without dependence on material culture, such as animal skins for clothing and fire for warmth. In essence, selected sites, such Olorgesailie, Olduvai, Omo, Turkana, St. Acheul, Torralba/Ambrona, Terra Amatta,

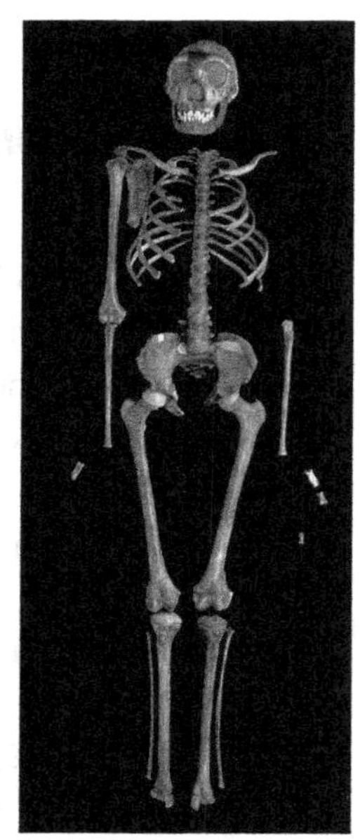

Figure 9.3. 'Homo erectus boy' discovered at Nariokotome, Kenya.

Gran Dolina, Ceprano, and Zhoukoudian show irrefutable evidence of the cultural behavior of *Homo erectus*.

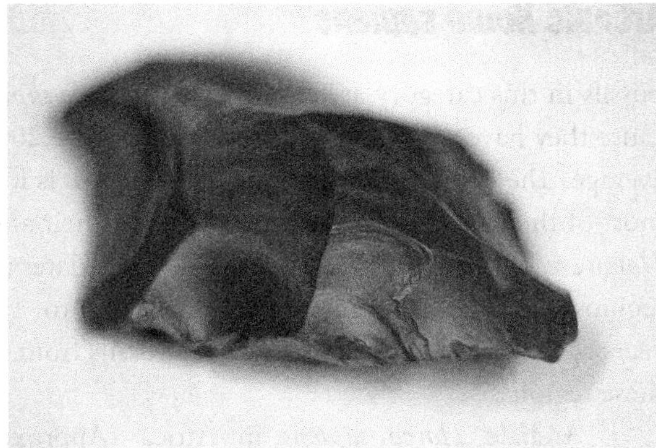

Figure 9.4. Acheulian hand ax.

TOOLS AND SUBSISTENCE

1. **Acheulian tool technology** (approximately 1.5 million years ago): hand ax (**biface**) (Figure 9.4)
   a. More efficient tool
   b. Required more skill to make than the Oldowan tools (suggests greater cognition)
   c. Multipurpose tool: possibly used for butchering large animals, digging up tubers, digging for water, defense against predators
2. **Caves/temporary shelters**
3. Hunting and gathering: evidence of *Homo erectus* fossils, fossils and tools, or tools found at the numerous sites listed above indicates that *Homo erectus*'s subsistence method was **hunting-gathering-scavenging**. Evidence from the archaeology is listed below:
   a. Large and small animal bones found (suggests that they hunted or scavenged large and small animals)
   b. Numerous stone tools found at **Olorgesailie** (probable animal butchering site)
   c. Fragmented animal bones (probable animal butchering)
   d. Dental evidence for **Homo erectus** (indicates significant meat in diet)
   e. Zhoukoudian (approximately 550,000 and 300,000 years ago): animal bones, tools, caves (possible living sites)
   f. Gathering (vegetables, fruits, nuts: 75 percent of the calories come from gathering)
4. *Homo erectus* and fire: very important for several reasons:
   a. Made possible the movement of Homo erectus into cold regions (higher latitudes)
   b. An important source of warmth and light
   c. Made "cooking" possible so that food was easier to chew
   d. Possibly used for tool manufacture
   e. Used to facilitate social interactions and scare away predators
   f. Zhoukoudian: first alleged evidence of controlled fire use

Speech and Language in *Homo erectus*?

Could *Homo erectus* speak? This is a difficult question to answer for two major reasons: (1) behaviors in general are not preserved in the fossil record, and (2) it is very difficult to discern brain structures from human skulls in general and fossil human skulls in particular. Nonetheless, some researchers (Begun and Walker 1993) have argued that the inferior frontal lobe where *Broca's area* is located is well developed in *Homo erectus*. Broca's area is involved in speech production and tool manufacture and use. Whether the Broca's area in *Homo erectus* is the same as the Broca's area in modern humans is unknown. So, could *Homo erectus* speak? It is possible that this species had some rudimentary form of communication, more sophisticated than that of the apes, but not as complex as modern human language. If cave paintings or sculptures were found in the *Homo erectus* archaeology, then we could argue that they used symbolic communication, which is a major characteristic of human language. To date, no cave paintings or sculptures have been found in the *Homo erectus* archaeological assemblage.

Homo erectus appeared and survived for almost 2 million years. They were a successful species because their big brain enabled them to exploit the environment, that is, hunt and eat meat; in turn, this changed their biology: it increased body size. In essence, *Homo erectus*' dependence on material culture made them human. But *Homo erectus*' continual march toward modernity would take an intermediate stage aptly called archaic *Homo sapiens*.

Archaic *Homo sapiens*

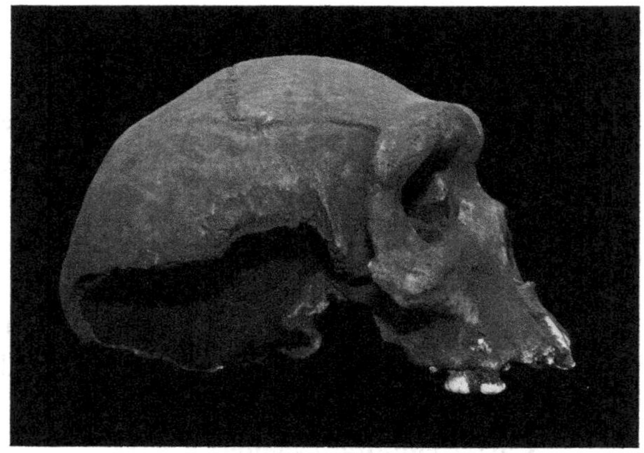

Figure 9.5. Archaic Homo sapiens Kabwe (Broken Hill) discovered in Zambia.

Fossils in this category are called archaic *Homo sapiens* because they have very large cranial capacities of 1,200 cc on average. The evidence for archaic *Homo sapiens* is found in most of the regions previously occupied by their ancestors *Homo erectus*: Africa, Asia, and Europe. Their dates in these regions span from approximately 350,000 to 130,000 years ago. Let us look at some selected fossils from each of these regions:

Archaic *Homo sapiens* in Africa (Approximately 300,000 to 200,000 Years Ago)

1. **Kabwe** (Broken Hill), Zambia (Figure 9.5):
   Skull:
   a. Sloping forehead
   b. Cranial capacity: 1,300 cc
   c. Large brow ridges
   d. Thick cranial bones
   e. No chin
2. **Bodo**, Ethiopia:
   Skull:
   a. Sloping forehead
   b. Massive face
   c. Large brow ridges
   d. Cut marks on forehead/eye sockets/cheekbones

Archaic *Homo sapiens* in Asia (Approximately 350,000 to 130,000 Years Ago)

1. **Ngandong** (Java), Indonesia:
   Skull:
   a. Sloping forehead
   b. Cranial capacity: 1,100 cc
   c. Large brow ridges
   d. Thick cranial bones

Archaic *Homo sapiens* in Europe (Approximately 350,000 to 130,000 Years Ago)

1. **Atapuerca**, Spain:
   Skull:
   a. Sloping forehead
   b. Cranial capacity: 1,125 cc
   c. Large brow ridges/facial projection
2. **Heidelberg** (Mauer), Germany:
   Mandible (jaw):
   a. Gracile
   b. Molar size similar to modern humans
   c. No chin
3. **Boxgrove**, England:
   Tibia (shin bone):
   a. Large bone

Technology: Acheulian tools

b. Petralona, Greece:

Skull:

a. Sloping forehead
b. Cranial capacity: 1,125 cc
c. Large brow ridges
d. Thick cranial bones
e. No chin

Face and skull are similar to Neanderthals

Lithic Technology

Archaic *Homo sapiens'* big brain enabled them to increase their level of sophistication in tool manufacture and tool use. They developed a new technique in making tools called the **Levallois technique**. This involved preparing the core by strategically knocking flakes off certain areas. Then the desired flakes were knocked off and saved for use. This technique produced sharp and symmetric tools. In short, archaic *Homo sapiens* had an excellent knowledge of flaking and were capable of precise tool making. Later, these behaviors would be exemplified in early modern *Homo sapiens.*

Summary of Key Ideas

1. *Pithecanthropus* was the original genus designation for *Homo erectus.*
2. *Homo erectus* is the first hominin species to be found outside of Africa. Dependence on material culture, that is, sophisticated stone tools for hunting and butchering animals, animal skins for clothing, and use of fire enabled *Homo erectus* to extend its geographical range.
3. *Homo erectus* were twice as large as australopithecines because of the high-quality nutrition of meat in their diet.
4. The *Homo erectus* archaeological assemblage suggests that this species was very human.
5. The Acheulian tool technology, that is, the hand ax, is associated with *Homo erectus.*
6 The *Homo erectus* species were hunters-gatherers-scavengers.
7. *Homo erectus* evolved into the big-brained archaic *Homo sapiens.*

Key Terms

Acheulian tool technology
Atapuerca
biface
Bodo
Boxgrove
Caves/temporary shelters
Eugene Dubois
Heidelberg
Homo erectus
Hunting-gathering-scavenging
Kabwe
Levallois technique
Nariokotome boy
Ngandong
Olorgesailie
Petralona
Pithecanthropus erectus
Zhoukoudian

Further Reading

Antón, S. C. 2003. A natural history of *Homo erectus. Yearbook of Physical Anthropology* 46: 126–70.

Rightmire, G. P. 1990. *The evolution of Homo erectus.* Cambridge, UK: Cambridge University Press.

Shipman, P., and P. Storm. 2002. Missing links: Eugene Dubois and the origins of paleoanthropology. *Evolutionary Anthropology* 11: 108–16.

INTERNET RESOURCES

Homo erectus
http://humanorigins.si.edu
This Web site is maintained by the Smithsonian Institution. It provides information and photographs of several *Homo erectus* specimens.

Thinking about the Hominin Family Tree
http://www.pbs.org/wgbh/evolution/humans/humankind/index.html
This Web site is maintained by PBS. It provides a species gallery of the best-known specimens representing each species.

Chimpanzee Communications and the Evolution of Human Language
http://www.sciencedaily.com/releases/2008/02/080228124415.htm
This Web site provides information on ape language studies and its implications for the evolution of human language.

Chapter 9: *Homo Erectus*

Study guide: By knowing the material below you will be well prepared for the exam. Don't forget to read the chapter and study your notes (this includes the chapter worksheets).

1. *Homo erectus* had a rich cultural life based on their tools and subsistence. Discuss their tool technology and their method of subsistence.
2. Why is *Homo erectus* species called the "first world travelers?"
3. Compared to australopithecines, *Homo erectus* was very large in body size. Why? Name four (4) cranial traits of *Homo erectus.*
4. Give the name of the person who found the first *Homo erectus* fossil. In what country and in what year did this individual discover the fossil?
5. The previous scientific name for *Homo erectus* was__________________________________.
6. What is the time span for archaic *Homo sapiens*? In what regions have they been found?
7. These fossils are placed in the category of archaic *Homo sapiens* because one key trait. Name that trait?
8. Be about to list the skeletal traits of the Narikotome boy

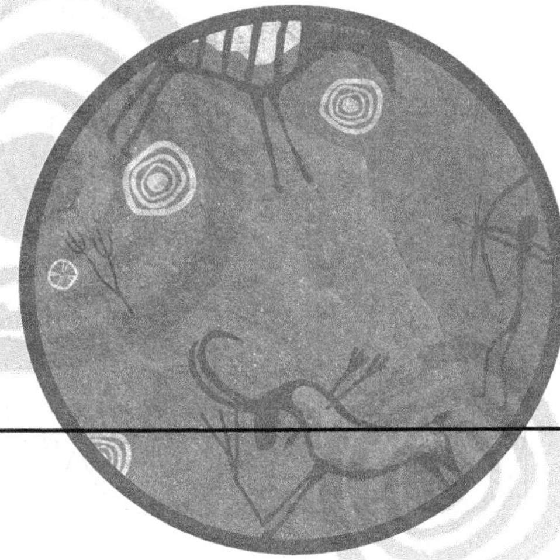

CHAPTER 10

The Origins of Modern Humans

The Hypotheses for Modern Human Origins

When and where did modern humans evolve? This question continues to generate tremendous debate in the literature and at human origins conferences. Did modern *Homo sapiens* evolve (from archaic *Homo sapiens*) in Africa, Europe, Asia, and Indonesia approximately 130,000 years ago? Or did modern *Homo sapiens* evolve only in Africa approximately 200,000 years ago? The multiregional continuity hypothesis answers and gives support to the first question, while the out of Africa hypothesis, also called the Eve hypothesis, answers and gives support to the second question.

MULTIREGIONAL CONTINUITY HYPOTHESIS

Recall from Chapter 8 that some of the *Homo erectus* groups left Africa approximately 2 million years ago, based on the fact that their fossils are also found in Europe, Asia, and Indonesia. According to the proponents of the **multiregional continuity hypothesis**, modern humans evolved in Africa, Europe, Asia, and Indonesia from *Homo erectus* (via archaic *Homo sapiens*) approximately 130,000 years ago (Wolpoff et al. 1994). They argue that the *Homo erectus* groups in Africa, Europe, Asia, and Indonesia were well adapted through natural selection to their respective environments. However, they emphasize that **gene flow** (interbreeding between *H. erectus* groups within and between the four big regions) kept the *Homo erectus* groups from **speciating** despite the different selective forces in each region. In Figure 10.1, the circles represent human groups adapting in their respective regions and the tight spacing of the circles represents gene flow across time and space. Small groups of animals have been known to speciate when reproductively isolated (no gene flow between groups) from the original population. In contrast, no human groups have been separated long enough to become different **species**. There have always been other groups nearby with whom gene flow through friendship (exogamy) or warfare (forced copulation) would take place. In summary, evolutionary change within the genus *Homo* is described as evolution within a single evolutionary **lineage** (**anagenesis**) from *Homo habilis* to modern *Homo sapiens* (see Figure 10.2).

OUT OF AFRICA HYPOTHESIS

The proponents of the **out of Africa hypothesis** argue that modern humans evolved from *Homo erectus* only in Africa approximately 200,000 years ago, and then some modern human groups migrated out of Africa generation after generation replacing (without interbreeding) the *Homo erectus* groups in Europe, Asia, and Indonesia (Cann et al. 1987; Stringer and Andrews 1988; Stringer and McKie 1996) (Figure 10.3). They emphasize that Neanderthals were actually a separate biological species (***Homo neanderthalensis***) and, therefore, not part of modern human lineage; they were an evolutionary dead end.

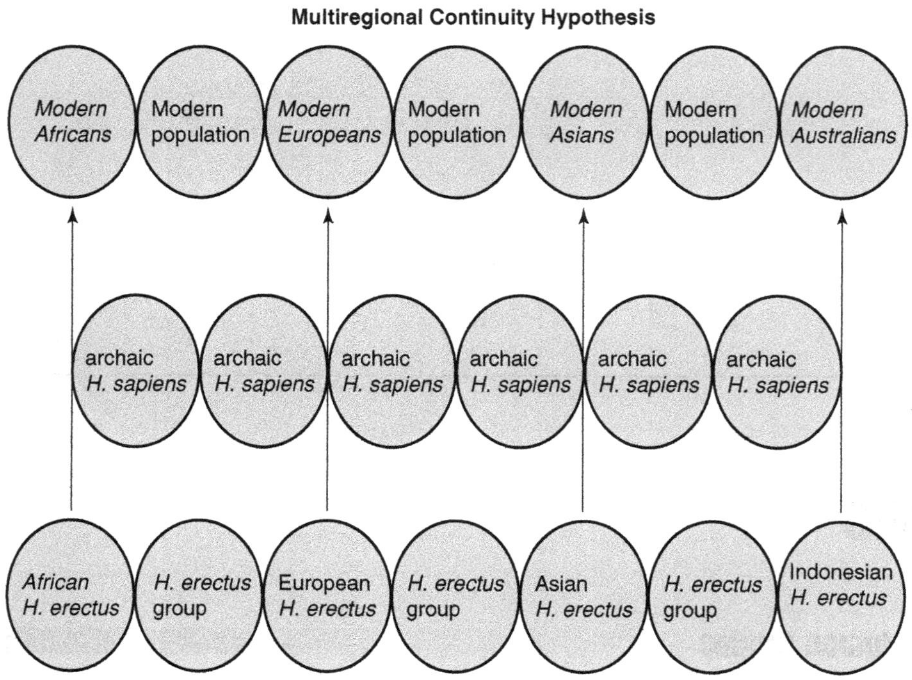

Figure 10.1. Multiregional evolution: a balance of selection and gene flow through time and space.

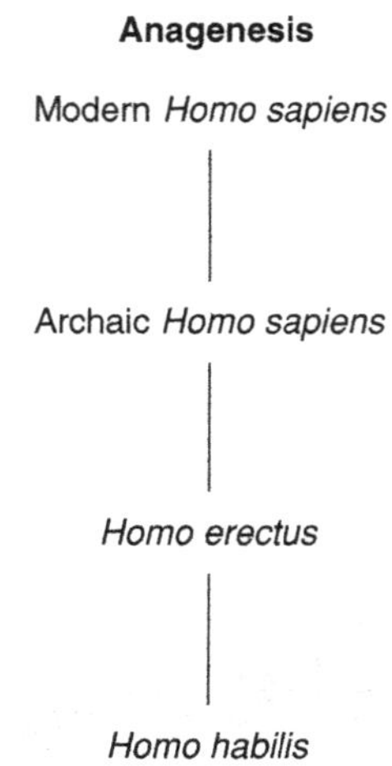

Figure 10.2. The evolution of modern humans within a single lineage.

The strongest evidence supporting the out of Africa hypothesis is **mitochondrial DNA (mtDNA)**. Males and females inherit an identical copy of their mother's mtDNA. Then the mtDNA is passed on, unchanged, to the next generation. But the mtDNA can only be tracked in females. Consequently, mtDNA can be used as a "**molecular clock**" to track the last common ancestor of populations (or two species) if there are several generations of daughters in a family. In the 1980s, placentas were collected from females living in the large geographical regions of Africa, Europe, Asia, and Indonesia. Mitochondrial DNA was extracted from these placentas and analyzed. Geneticists discovered that the African mtDNA was more variable compared to the European and Asian mtDNAs. At this point, they made the assumption that this variability indicated more mutations. As a result of these numerous mutations, the "logical" assumption was that the African mtDNA was older, simply because it takes time to accumulate numerous mutations. From these assumptions, they concluded that modern humans evolved from a female who lived in a group in Africa 200,000 years ago—hence the title "the **Eve hypothesis**."

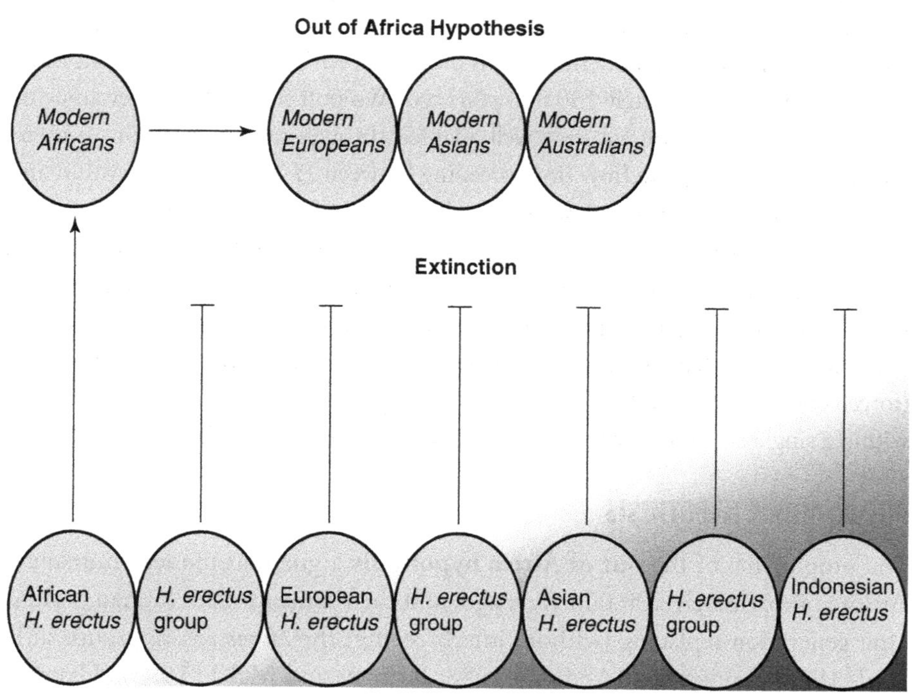

Figure 10.3. Out of Africa: total replacement of archaic Homo species outside of Africa (by early modern Africans) without gene flow.

There is no strong fossil evidence supporting the out of Africa hypothesis. In summary, evolutionary change within the genus *Homo* is described as branching evolution (**cladogenesis**), where the African *Homo erectus* evolved into a new species (*Homo sapiens*) and the *Homo erectus* groups outside of Africa were evolutionary dead ends (separate biological species) (see Figure 10.4).

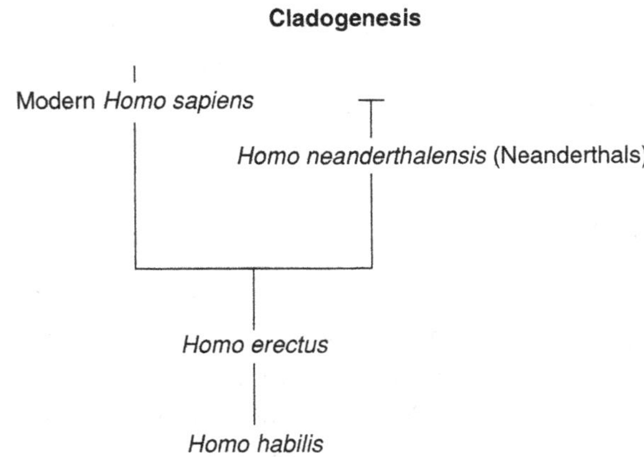

Figure 10.4. The evolution of modern humans through speciation (branching evolution).

ASSIMILATION HYPOTHESIS

Since the proponents of both the multiregional continuity and out of Africa hypotheses cannot disprove each other's arguments, a compromise was proposed. Proponents of the **assimilation hypothesis** proposed that the transition from *Homo erectus* to modern *Homo sapiens* did occur in Africa, but these changes then spread to the other archaic groups outside of Africa through gene flow. In essence, the genes of non-African archaic groups were assimilated into the gene pool of the expanding modern African groups (rather than no interbreeding or extinction).

New DNA sequencing techniques have provided additional support for the assimilation hypothesis. Richard Green et al. 2006, using new methods to sequence degraded fragments of ancient DNA, spliced together a composite sequence from three female Neanderthals who lived in Croatia 38,000 to 44,000 years ago to construct about two-thirds of the entire Neanderthal genome. Reading the sequence, the researchers concluded that modern Europeans and Asians—but not Africans—have inherited between 1% to 4% of their genes from Neanderthals. Apparently, Neanderthals in the Levant (Middle East) interbred with the modern Africans after they entered the Levant at least 80,000 years ago. Subsequently, the modern humans that left the Levant spreading to Europe and Asia had 1% to 4% Neanderthal genes.

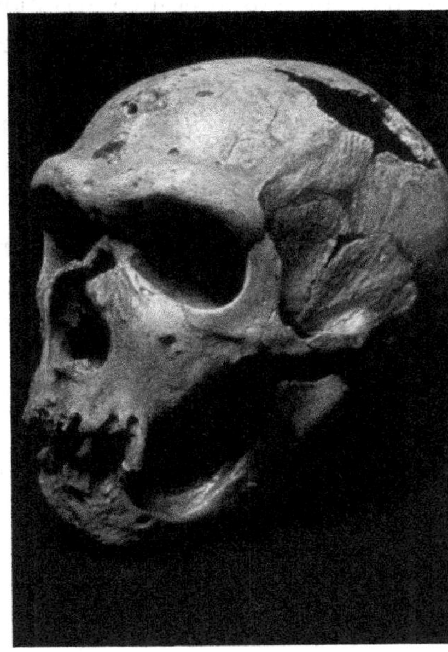

Figure 10.5. An elderly Neanderthal man discovered at Chapelle-aux-Saint, France. John Reader/Photo Researchers, Inc.

Neanderthals (130,000 to 24,000 years ago)

In 1856, fossilized bones were found in a cave near Düsseldorf, Germany. The cave is located in the Neander Valley (in German, valley is *thal*), hence the name **Neanderthal**. Since most people at this time believed that "man" was created in the image of God and, therefore, could not have prehistoric ancestors, these Neanderthal bones—a long, big-browed skull and bowed, thick limb bones—were viewed as belonging to a modern human afflicted with a bone disease. While this pathological hypothesis had no lasting effect on the perception of Neanderthals, another hypothesis will have the effect of relegating Neanderthals to an evolutionary side branch. In the early 1900s, a detailed analysis of a Neanderthal skeleton was undertaken by the prominent French paleoanthropologist Marcellin Boule (1861–1942). Based on Boule's analysis of the bones, the Neanderthal walked with a bent-kneed gait; he was not able to speak; he was stupid and too primitive to have evolved into modern humans (Figure 10.5). However, Boule did not realize that this skeleton belonged to an elderly Neanderthal man with several bone pathologies. Nonetheless, Boule's description would be very influential in coloring later scientists' and the public's perception of Neanderthal behavior and place in human evolution. In essence, the name "Neanderthal" carries with it much negative baggage, synonymous with dumb and stupid. Despite this fact, we will continue to use the name Neanderthal and avoid the contentious scientific taxonomic designations *Homo neanderthalensis* (indicating a different species not ancestral to modern humans) and *Homo sapiens neanderthalensis* (indicating an extinct race of archaic humans ancestral to modern humans).

GEOGRAPHY

Neanderthals appear in the fossil record at approximately 130,000 years ago and were gone (or were genetically swamped by other modern human groups) by 24,000 years ago. Based on the archaeological sites in Europe and Asia, Neanderthals ranged from Spain to Uzbekistan. Specifically, Neanderthal sites are located in the following regions: France (Iberia, Zafarraya, Le Moustier, La Ferrassie, St. Césaire, La Chapelle-Saints, Spy); Germany (Feldhofer); Croatia (Krapina); Ukraine (Kiik-Koba); Uzbekistan (Teshik Tash); Iraq (Shanidar); and Israel (Amud, Kebara, Skhul, Tabun, Zuttiyeh) (Campbell et al. 2006; Day 1965; Larsen 2008; Relethford 2005).

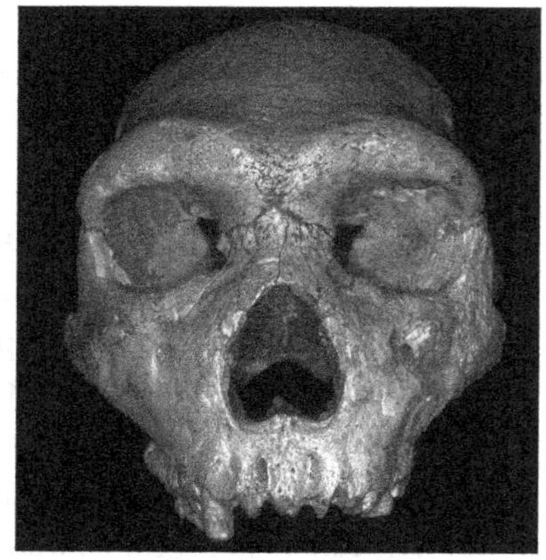

Figure 10.6. Classic Neanderthal features: large brow ridges and large nasal opening.

Milford H. Wolpoff, University of Michigan

NEANDERTHAL BIOLOGY

Generally, Neanderthals were a cold-adapted group of early humans. They were stocky with large body size and short, thick limbs. This morphology minimizes surface area in order to conserve heat. According to the **Bergmann** and **Allen rules** of adaptation, mammal populations in cold climates will have large body sizes and short limbs, which minimizes the surface area to volume. This aids in heat conservation. In contrast, mammal populations in hot climates will have small body sizes and long limbs, which maximizes the surface area to volume. This aids in heat dissipation. It is interesting that the Neanderthals in the Middle and Near East were less robust than the European Neanderthals. Let us now look at some of the common features of the cranial and postcranial that are associated with Neanderthals (Figure 10.6).

Skull/face:Cranial capacity (averaging 1,500 cc)

1. Large brow ridges
2. Sloping forehead
3. Mid-facial projection (indicates massive nasal cavities to warm and moisten cold, dry air)
4. Space behind third molars (possibly caused by mid-face projection)
5. Very large incisors
6. No chin
7. Occipital bun (back of skull has a bump on it)

The Neanderthal postcranial, such as the pelvis, ribs, carpals (wrist bones), tarsals (ankle bones), metacarpals (hand bones), metatarsals (foot bones), and phalanges (fingers and toes) are all large and thick. In addition, their humeri (upper arm bones), ulnae/radii (lower arm bones), femora, and tibiae are all thick, short, and bowing due to large muscle attachments. These cranial and postcranial traits indicate that Neanderthals were very strong people, a cold-adapted race of human beings (Figure 10.7).

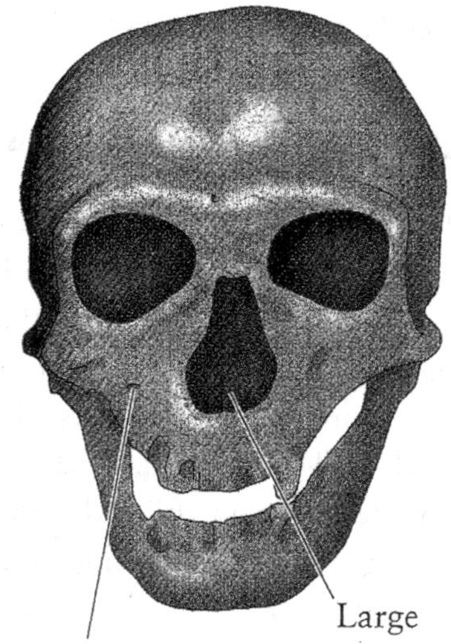

Figure 10.7. The large nasal opening correlates with adaptation to a cold, dry environment.

"Figure 11.24", from *Our Origins* by Clark Spencer Larsen. Copyright (c) 2008 by W. W. Norton & Company, Inc. Used by permission of W. W. Norton & Company, Inc.

NEANDERTHAL DNA

In 1997, molecular anthropologists were successful in extracting a sequence consisting of 378 mtDNA bases from a Neanderthal fossil dated at approximately 40,000 years old (Krings et al. 1997). Mitochondrial DNA has become very popular in evolutionary studies because it can be used when biological materials are either ancient or degraded. Continuing, the analysis showed that there were 27 base pair differences between Neanderthal mtDNA and living human DNA. The difference between any two humans is only 8 base pairs. On the surface, this is strong evidence that Neanderthals are not early humans or not within the human lineage. But these results are preliminary, and there are probably a myriad of changes that DNA experiences over 40,000 years. In essence, the large genetic differences between Neanderthals and living humans might be a function of time (changes in the mtDNA sequences over time) and not necessarily phylogenetic or species differences.

Recently, another team of molecular anthropologists (Green et al. 2006) has successfully extracted DNA from a 38,000 year old Croatian Neanderthal specimen, and they are now trying to sequence the entire Neanderthal genome (Hall 2008). So far, they have found that there is no difference between Neanderthals and modern humans for most of the sequence. Nonetheless, they have found a few important differences suggesting a divergence time of approximately 700,000 years between the *Homo erectus* populations leading to modern humans and the *Homo erectus* populations leading to Neanderthals. The researchers have identified 15 regions of interest that differ between humans and Neanderthals including genes that are important in cognitive and skeletal development. In summary, much more work needs to be done to understand Neanderthal and modern human origins.

CULTURE AND BEHAVIOR

The stone tool technology of Neanderthals is known as the **Mousterian complex**. It is very similar to the technology of archaic *Homo sapiens*—the prepared tool technique (see Chapter 8). But Neanderthals added more variety in their tool kit including the stabbing spear. These tools combined with Neanderthals' robust body size enabled them to become expert hunters. The Neanderthal fossil record shows butchering sites containing bones of large animals. Neanderthals also ate plant food, including cooked plants. This statement is based on plaque stuck to the teeth of three Neanderthals who died approximately 35,000 years ago (Henry et al. 2011). In addition, the Neanderthal bones themselves have many healed injuries, suggesting that the prey animals did not go down easily. Anthropologists Thomas Berger and Erik Trinkaus (1995) compared the injuries on Neanderthal bones as similar to those of rodeo riders.

Since most of the Neanderthal fossils are found buried, we can assume that they buried their dead. Also, there is evidence (some Neanderthals skeletons age 50–55 years old with healed injuries) suggesting that they took care of the injured and elderly. Both of these behaviors are very complex; they add intelligence to the traditionally "primitive" image of Neanderthals.

SPEECH AND LANGUAGE IN NEANDERTHALS

Did Neanderthals have language? We will never know for sure. However, using their cultural and behavioral sophistication as indirect evidence, it is difficult to argue that they did not have language abilities. Furthermore, analysis of the first Neanderthal hyoid bone showed that it is identical in size and shape to the hyoid bone of modern humans. The **hyoid bone** is one of the important structures in the throat that enables humans to make the sounds necessary for speech. In 2007, a version of the **FOXP2** gene (the same version in modern humans) was found in Neanderthal DNA. The FOXP2 has been identified in all mammals and may have mutated twice in evolutionary history, coinciding with the origins of speech and language ability.

RECENT DISCOVERY: THE DENISOVANS (APPROXIMATELY 50,000 TO 30,000 YEARS AGO)

In 2008, Russian archaeologists Mikhail Shunkov and Anatoly Derevianko from the Institute of Archaeology and Ethnology of Novosibirsk discovered a cave in southern Siberia, 350 kilometers north of the Russian border with both Kazakstan and Mongolia. Now the discoveries have shifted the arguments on the origins of modern humans to Asia because there is evidence that place these Denisovans—the name of the cave where the skeletal remains and artifacts were recovered—with modern humans and Neanderthals.

Since no cranial remains were recovered, the evolutionary relationships of these fossils are based solely on genetics and no taxonomic designation can be rendered because of this fact and the current paucity of skeletal remains.

The skeletal remains that were found in layer 11 of the cave were a molar, a finger bone, and a toe bone (figure 10.8a). Based on nuclear DNA sequencing, the finger bone and molar were very different from modern humans and any archaic humans (Reich et al. 2010; Krause et al. 2010). The mitochondrial DNA from the toe bone indicated Neanderthal. Derevianko and colleagues present a scenario in which the Denisovans were in the cave about 50,000 years ago (radiocarbon dating of seven animal bones with cut marks from layer 11 provides dates of 50,000 years ago or older), Neanderthals came in briefly about 40,000 years ago, and modern humans followed (Gibbons 2011).

The above scenario is based on the genetic data because 2% to 7% of modern European and Asian DNA, apparently, came from the archaic Denisovans and Neanderthals (Reich et al. 2010; Krause et al. 2010). Specifically, the researchers examined HLA (Human Leucocyte Antigen System which helps the immune system recognize pathogens) alleles in one Denisovan and several Neanderthals and found several ancient HLA variants that appear in the genomes of living Asians and Europeans. They also found that Melanesians and Australian Aborigines have inherited 5% of Denisovan DNA leading to the conclusion that the Denisovans and the ancestors of Melanesians must have interbred before 40,000 to 60,000 years ago when Australian Aborigines first settled Australia.

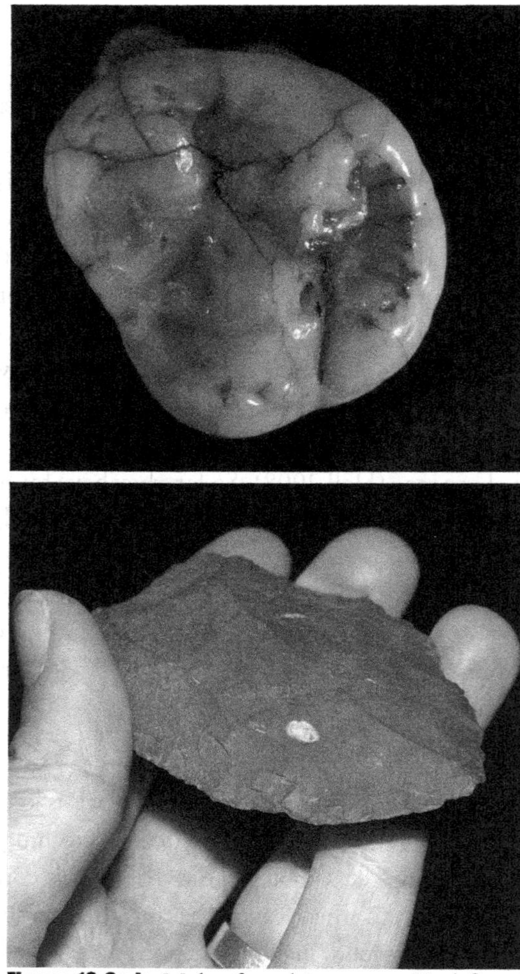

Figure 10.8a,b. Molar found in a cave, Tool found in a cave

Dr. Bence Viola, Max-Planck-Institute of Evolutionary Anthropology.

The archaeology consists of two different types of local cultures of artifacts: Late Paleolithic and Upper Paleolithic. Excavated from the cave were Mousterian stone points and scrapers; stone bladelets used on spears; pendants made of teeth of fox, bison, and deer; and a stone bracelet (figure 10.8b).

In conclusion, work is continuing in archaeology, genetics, and skeletal anatomy to illuminate the Denisovans and how they relate to us.

Other Early Moderns

ASIA AND EUROPE (APPROXIMATELY 90,000 TO 15,000 YEARS AGO)

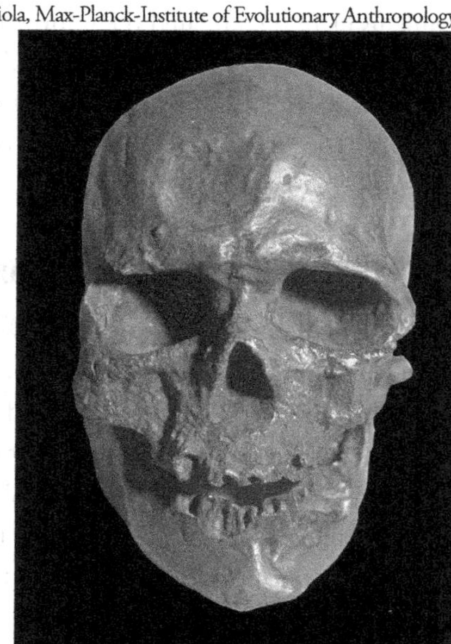

Figure 10.9. Cro-Magnon: the archetype fossil for modern humans.

The **Qafzeh** and **Skhul** fossils, found in Israel, are called **Proto-Cro-Magnons.** They are believed to be ancestors of later Cro-Magnons of Europe. The **Cro-Magnons**, found in France, are frequently used as the archetype fossils for modern humans. They have well-rounded skulls, high foreheads, small brows, non-protruding faces, and strong chins (Figure 10.9). These characteristics fit some modern human populations but not all. Two modern populations in which the Cro-Magnon traits do not fit are Australian Aborigines and sub-Saharan Africans. This indicates that "modern features" are different in different regions of the world.

LITHIC TECHNOLOGY

The tool technology associated with the Cro-Magnons is called the **Aurignacian complex**. It contains a higher frequency of smaller tools, bone tools, and projectile points:

- **Bone tools** (bone needles)
- **Blades** (stone tools that are longer than they are wide)
- **Burins** (used for carving)
- Fine stone points
- **Barbed harpoons**
- Fishing nets

Furthermore, **cave paintings** and **carved figurines** (figures of females with exaggerated breasts and buttocks) are a major part of this modern archaeological assemblage (Figure 10.10). Why did early humans paint and carve images of themselves and animals? We can only speculate. Painting the image of an animal on the cave wall might result in a successful hunt, or success in birthing in the case of the carved figurines. Whatever the reason, early humans were expressing themselves symbolically.

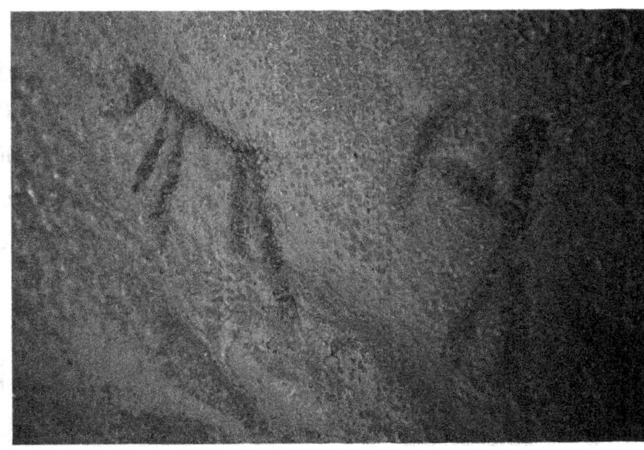

Figure 10.10. Cave paintings: Upper Paleolithic.

The Cultural Revolution: The Shift to Agriculture

As global temperatures continue to rise in the last 25,000 years, most humans were forced to find a new way of life. Forests advanced, while the savanna/woodland retreated. The large prey animals that were adapted to temperate climates, such as the woolly mammoth, migrated northward and some hunters-and-gatherers followed them. These human groups eventually entered the New World, approximately 15,000 years ago, via a land bridge joining Asia and North America. The remaining human groups began collecting seeds and capturing large animals. Eventually, this would lead to **plant** and **animal domestication**. Later, the shift to full-time **agriculture** would take place.

Agriculture over time leads to **sedentism**. When you plant your crops, you cannot just leave. You have to stay and cultivate these crops. More people are surviving because agriculture can feed many people. Soon, there are towns, cities, states, and population density. People are living next to their livestock and garbage. Over time, an optimal environment forms for disease epidemics.

Peopling of the "New World"

As stated in the last section of this chapter, some human groups followed the migrations of the large, cold adapted animals to north. Eventually, these animals would cross over into North America via a land bridge called ***Beringia*** joining eastern Siberia to present-day Alaska, and these human groups or hunters-and-gatherers would follow (Figure 10.11). An initial migration into North America that began approximately 15,000 years ago would continue into Central and South America (Figure 10.11). These ancient hunters-and-gatherers would be anointed with the name ***Paleoindians***.

What is the evidence for this 15,000 year date? The evidence comes from archaeology and genetics. As far as the archaeology is concerned, the earliest well-documented archaeological site of habitation and material culture is in North America (Clovis, New Mexico), dated at approximately 11,000 years ago. Naturally, the material culture dated at this time is called the **Clovis tradition**. Also, thousands of miles south in central Chile, an archaeological site called Monte Verde dates firmly to 12,000 years ago (Smith 1995). Based on this 12,000 year date, could the first migration of humans into the Americas have begun earlier than 15,000 years ago? Most archaeologists and geologists are skeptical because during this last glaciation, there was solid ice blocking the way into North America. In other words, there were no ice-free paths into North America before 20,000 years ago. The genetic data also indicates a date less than 20,000 years ago. Specifically, genetic dating based on mutation rates of mtDNA and Y-chromosomes indicate that the peopling of the New World took place approximately 15,000 years ago.

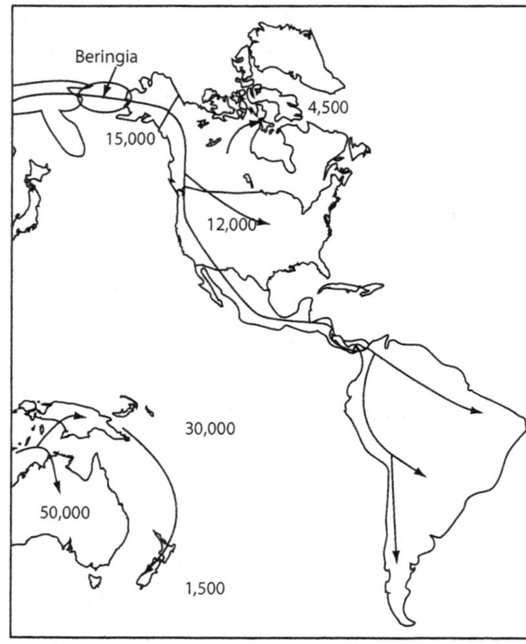

Figure 10.11. Map showing migrations into North and South America via the Beringia beginning 15,000 years ago.

Asian Ancestry for Native Americans: The Dental and Genetic Evidence

There is a high frequency of shovel-shaped incisors in Northeast Asians and Native Americans. This similarity suggests an ancestral link between Northeast Asians and Native Americans. Even more compelling are the numerous genetic evidence, of which two major analyses are mentioned above. In an early study, 120 allele frequencies for red and white blood cell polymorphisms in Native Americans and other populations were analyzed for genetic distances. The lower the genetic distance, the greater the genetic similarity. The results showed that Native Americans were more similar genetically to Northeast Asians than to other populations (Cavalli-Sforza et al. 1994).

In more recent analyses using mtDNA **haplogroups** (as discussed at the beginning of this chapter and in Chapter 3, mtDNA is found in the cytoplasm and can be used as a molecular clock because it is passed on unchanged from female to female), 95% of living and ancient Native Americans throughout North and South America belong to one of four haplogroups labeled A, B, C, D (Schurr 2000; Relethford 2005). Interestingly, these haplogroups are found in Northeast Asian populations (Siberia and East Asia) and are absent in other parts of the world. Consequently, Native Americans are the descendants of migrants from Northeast Asia.

Furthermore, did these migrants from Northeast Asia cross the Beringia once as a huge population, or were there separate migration events from different regions of Asia? As mentioned above, the mtDNA haplogroups of Native Americans are found in Siberia and East Asia, suggesting two migration events. In addition, analysis of Y-chromosome **haplotypes**, labeled 1C and 1F, found in high frequency in Native Americans are also distributed in different regions in Asia (Karafet et al. 1999; Relethford 2005). This is additional evidence suggesting two separate migration events.

Paleoindians: Culture and Behavior

The paleoindians were experienced hunters. They are associated with the Clovis and Folsom stone tool cultures in North America and comparable tools in South America (Figure 10.12). Such points were large, bifacial, and fluted (or grooved) at the bottom in order to attach a spear. In essence, this technology was designed to hunt **megafauna** (large game animals), such as the woolly mammoth, bison, and bears, to name a few. The paleoindians, compared to their immediate and distant ancestors in the Old World, made a greater impact on the environment. They were dangerous and were probably responsible for the extinction of many large game animals, particularly the woolly mammoth. Nonetheless, their lifestyle would persist until contact with Europeans and western civilization.

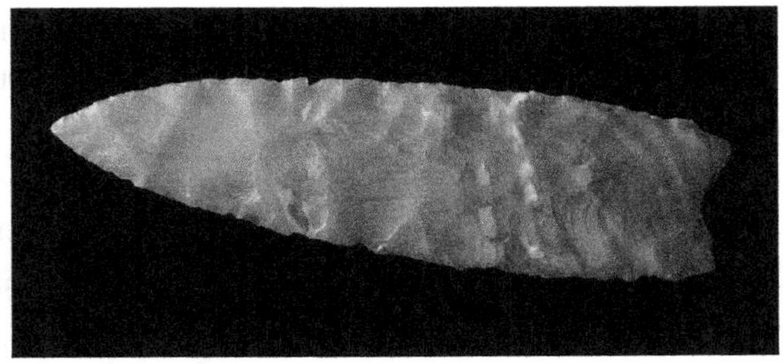

Figure 10.12. A Clovis spear point used between 12,000 and 11,000 years ago. © Warren Morgan/CORBIS

Summary of Key Ideas

1. The multiregional continuity hypothesis states that modern humans evolved from *Homo erectus* in several regions of the Old World approximately 130,000 years ago with a balance of gene flow and natural selection.
2. The mode of evolution within the multiregional continuity hypothesis is anagenesis or evolution within a single lineage.
3. The out of Africa hypothesis states that modern humans evolved from *Homo erectus* only in Africa, approximately 200,000 years ago, then some groups moved out and replaced/outcompeted (without interbreeding) the *Homo erectus* groups outside of Africa.
4. The mode of evolution within the out of Africa hypothesis is cladogenesis or branching evolution, where Neanderthals are a separate branch.
5. The strongest evidence supporting the out of Africa hypothesis is mitochondrial DNA (mtDNA).
6. The assimilation hypothesis states that modern humans did evolve from *Homo erectus* in Africa approximately 200,000 years ago. But the modern changes spread throughout the Old World outside of Africa through gene flow.
7. Marcellin Boule's 1901 description of a complete Neanderthal skeleton as being bent-kneed, apelike, stupid, and so on would be very influential in coloring later scientists' and the public's perception of Neanderthals' behavior and place in human evolution.
8. The Neanderthals were a cold-adapted race of humans with robust bodies and muscular, short limbs.
9. Analysis of Neanderthal DNA showed that there were 27 base pair differences between Neanderthal mtDNA and living human DNA.
10. The tool technology of the Neanderthals is the Mousterian complex.
11. The Neanderthals' major cultural method of subsistence was hunting.
12. Neanderthals buried their dead and possibly took care of the elderly.
13. Modern features are different in different regions.
14. Bone tools, blades (stone tools that are larger than they are wide), projectile points, and barbed harpoons are part of the Aurignacian complex.
15. Cave art and carved figurines indicate that early humans were expressing themselves symbolically.
16. Increase in global temperatures in the last 25,000 years leads to a shift from hunting-and-gathering to agriculture.
17. The Beringia joins eastern Siberia to present-day Alaska.
18. An initial migration into North America began approximately 15,000 years ago.

19. The Clovis tradition is the earliest Native American culture of North America. The stone technology (projectile) is large, bifacial, and fluted.
20. Genetic dating based on mutation rates of mtDNA and Y-chromosomes indicate that the peopling of the New World took place approximately 15,000 years ago.
21. There is a high frequency of shovel-shaped incisors in Asians and Native Americans, suggesting an ancestral-descendant relationship.
22. Most of the living and ancient Native Americans throughout North and South America belong to one of the four haplogroups labeled A, B, C, D that are also found in Northeast Asian populations.
23. Paleoindians were probably responsible for the extinction of the woolly mammoth.

Key Terms

Agriculture
Allen rule
Anagenesis
Animal domestication
Assimilation hypothesis
Aurignacian complex
Barbed harpoons
Bergmann rule
Beringia
Blades
Bone tools
Burins
Carved figurines
Cave paintings
Cladogenesis
Clovis Tradition
Cro-Magnons
FOXP2 gene
Gene flow
Haplogroups
Haplotypes
Homo neanderthalensis
Homo sapiens neanderthalensis
Hyoid bone
Lineage
Marcellin Boule
Megafauna
Mitochondrial DNA (mtDNA)
Molecular clock
Mousterian complex
Multiregional continuity hypothesis
Neanderthal
Out of Africa hypothesis (Eve hypothesis)
Paleoindians
Plant domestication
Proto-Cro-Magnons
Qaf Sedentism zeh
Skhul
Speciation
Species

Further Readings

Diamond, J. *Guns, germs, and steel.* New York: W. W. Norton.

O'Rourke, D. H., M. G. Hayes, and S. W. Carlyle. 2000. Ancient DNA studies in physical anthropology. *Annual Review of Anthropology* 29: 217-242.

Steele, D. G. and J. F. Powell. 1993. Paleobiology of the first Americans. *Evolutionary Anthropology* 2: 138-146.

Stringer, C., and C. Gamble. 1993. *In search of the Neandertals.* New York: Thames & Hudson.

Tattersall, I. 1995. *The last Neanderthal.* New York: Macmillan.

Trinkaus, E., and P. Shipman. 1993. *The Neandertals.* New York: Knopf.

INTERNET RESOURCES

Homo neanderthalensis
http://humanorigins.si.edu
This Web site is maintained by the Smithsonian Institution. It provides information and photographs of Neanderthal fossils.
Comparing Morphology and Genetic Evidence
http://www.pbs.org/wgbh/nova/neanderthals/skulls.html
http://www.pbs.org/wgbh/nova/neanderthals/mtdna.html
These Web sites provides photographs of Neanderthal skulls and information on the genetic relationship between modern humans and Neanderthals.
The Origin of Modern Humans: Multiregional and Replacement Hypotheses
http://evolution.suite101.com/article.cfm/out-of-africa-vs-multiregional-hypothesis
This Web site was created by biologist Michael Roberts. It examines the two competing hypotheses for modern human origins.

Chapter 10: The Origins of Modern Humans

Study guide: By knowing the material below you will be well prepared for the exam. Don't forget to read the chapter and study your notes (this includes the chapter worksheets).

1. Discuss the Neanderthals. Include: 1) their geography (where they lived); 2) their skeletal characteristics; and 3) culture. Based on the evidence, are they ancestral to modern humans? Discuss this evidence.
2. Discuss the Out of Africa model for modern humans origins
3. Discuss the Multiregional model for modern human origins
4. Discuss the Assimilation model for modern human origins
5. The Out of African and Assimilation models are identical. On what point do these models differ?
6. On what point do the Multiregional and Assimilation models agree?
7. Discuss recent genetic evidence that reinforces the Assimilation model
8. Distinguish between the taxonomic designations *Homo neanderthalensis* and *Homo sapiens neanderthalensis*?
9. Who are the Denisovans? How are they related to modern humans?
10. Who are the Cro-Magnons? What tool technology did they use?
11. Why did humans 12,000 years ago shift to agriculture? What was the bio-cultural impact of this cultural revolution?
12. What is the approximate date from the first migration into North America? Who (what species) were these migrants? Define Beringia.
13. What is the dental trait that is in high frequency in Native American populations?
14. Discuss the genetic evidence that puts Native Americans ancestry in Northeast Asia.
15. Discuss the culture of the Paleoindians

NAME ______________________ **DATE** ______________

Chapter 10: The Origins of Modern Humans

HOME WORK EXERCISES (10PTS.)

The goal of the home work exercises is assess your knowledge of the concepts and ideas presented in the chapter. Please answer the questions in each section as you go. When class lecture on this chapter is completed, carefully tear the home work exercises from the textbook and turn it in. **[Bonus pts. = 10]**

Section 10.1: The Hypotheses for Modern Human Origins

1. Discuss the Multiregional Continuity Hypothesis. What is the key evidence used to support this hypothesis?

2. Discuss the Out of Africa Hypothesis. What is the key evidence used to support this hypothesis?

3. Discuss the Assimilation Hypothesis. How does this hypothesis differ from the Out of Africa Hypothesis? How is it similar to the Multiregional Hypothesis?

4. Discuss recent genetic evidence that reinforces the Assimilation Hypothesis

Section 10.2: Neanderthals (130,000 to 24,000 years ago)

1. After reading the short history of Neanderthals in the text, why does the name "Neanderthal" portray such a negative image?

2. Distinguish between the taxonomic designations *Homo neanderthalensis* and *Homo sapiens neanderthalensis*?

3. Discuss the Neanderthals. Include: 1) their geography (where they lived); 2) their skeletal characteristics; and 3) culture. Based on the evidence, are they ancestral to modern humans? Discuss this evidence.

4. Who are the Denisovans? How are they related to modern humans?

Section 10.3: Other Modern Humans

1. Who are the Cro-Magnons? What tool technology did they use?

Section 10.4: The Cultural Revolution: The Shift to Agriculture

1. Why did humans 12,000 years ago shift to agriculture? What was the bio-cultural impact of this cultural revolution?

Section 10.5: Peopling of the "New World"

1. What is the approximate date from the first migration into North America? Who (what species) were these migrants? Define Beringia.

Section 10.6: Asian Ancestry for Native Americans: The Dental and Genetic Evidence

1. What is the dental trait that is in high frequency in Native American populations?

2. Discuss the genetic evidence that puts Native Americans ancestry in Northeast Asia.

Section 10.7: PaleoIndians: Culture and Behavior

1. Discuss the culture of the Paleoindians

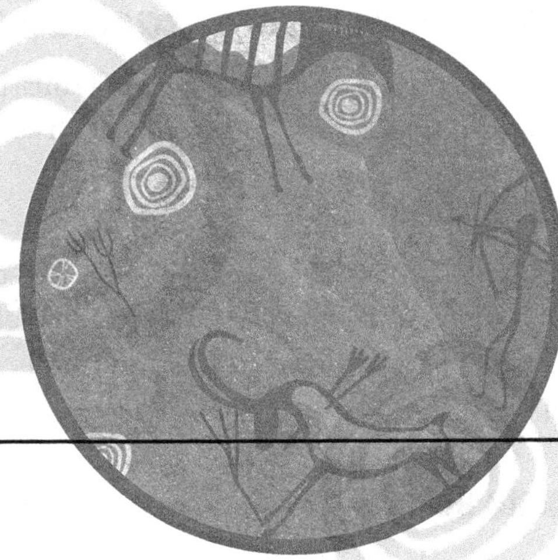

CHAPTER 11

Modern Humans: Climate, High Altitude, and Cultural Adaptation

Human Adaptation

When we look at humans, we see a highly variable biological species. This is because modern humans and their prehistoric and historic ancestors have been adapting for hundreds of thousands of years in Africa, Europe, and Asia, so that populations today have inherited their ancestors' traits. Despite the fact that humans do tan (except for albinos), general skin color, hair form, and nose shape will not change in our lifetime, regardless of changing environments. But body size and shape will change in our lifetime, in respective environments. This is the biocultural interaction discussed in Chapter 1.

We have already discussed two important examples of human adaptation, such as the sickle cell trait in malarial environments and the Bergman-Allen rules (mammalian body size and shape) of adaptation in hot and cold environments in chapters 3 and 10. Let us now elaborate on **climate adaptation**, then focus on **high altitude adaptation,** and the impact of nutrition and disease on human biology.

Types of Adaptation

GENETIC ADAPTATION

Human skin color (race/ancestry) is an example of **genetic adaptation** because it has been inherited within a population from generation to generation, over hundreds of thousands of years and it cannot be reversed.

DEVELOPMENTAL ADAPTATION

If a baby born in Bloomsburg, Pennsylvania (low altitude) was adopted by a native couple from Tibet or the Peruvian Andes (both at high altitude), that baby will develop an efficient oxygen transport system similar to (but not identical with) the native inhabitants. Since this **developmental adaptation** occurs during the individual's growth and development, it is irreversible.

PHYSIOLOGICAL ADAPTATION 1 (ACCLIMATION)

When a person is hot, he/she begins to sweat. Or, exposure to sunlight for extended periods of time results in tanning. These are examples of **physiological adaptation** or short-term responses to an environmental stress. It is reversible.

PHYSIOLOGICAL ADAPTATION 2 (ACCLIMATIZATION)

If a person from Bloomsburg, Pennsylvania moves to Tibet, for several months there will be an increase in red blood cell production in order to maintain sufficient oxygen levels in this low oxygen environment. It is reversible.

CULTURAL ADAPTATION

In cold weather, we wear warm clothing. Or to build a house in the forest, we would use an axe or a chainsaw to clear a section of the forest. These cultural tools (technology) are examples of cultural adaptations. And the definitive beginnings of **cultural adaptation** in the human lineage go back to *Homo erectus* during the Pleistocene (see Chapter 9).

Climate Adaptation

COLD STRESS

When the body temperature falls rapidly below 98.6 degrees F, a condition called **hypothermia** occurs. If this condition is not quickly reversed, then the individual may lose a limb or die. The goal is to reduce heat loss. In less extreme circumstances, the physiological responses are shivering, which increases the metabolic rate, then the narrowing (**vasoconstriction**) and widening (**vasodilation**) of blood vessels. Vasoconstriction *reduces* blood flow and *prevents* heat loss from the core of the body, and vasodilation *increases* blood flow and *facilitates* heat loss from the core of the body. Vasodilation is needed in a cold environment because prolonged vasoconstriction—blood flow is reduced to the extremities—can cause frostbite and loss of extremities. Fortunately, this cycle of vasoconstriction and vasodilation becomes more frequent and less extreme providing more efficient adaptation (Figure 11.1) [Frisancho 1993].

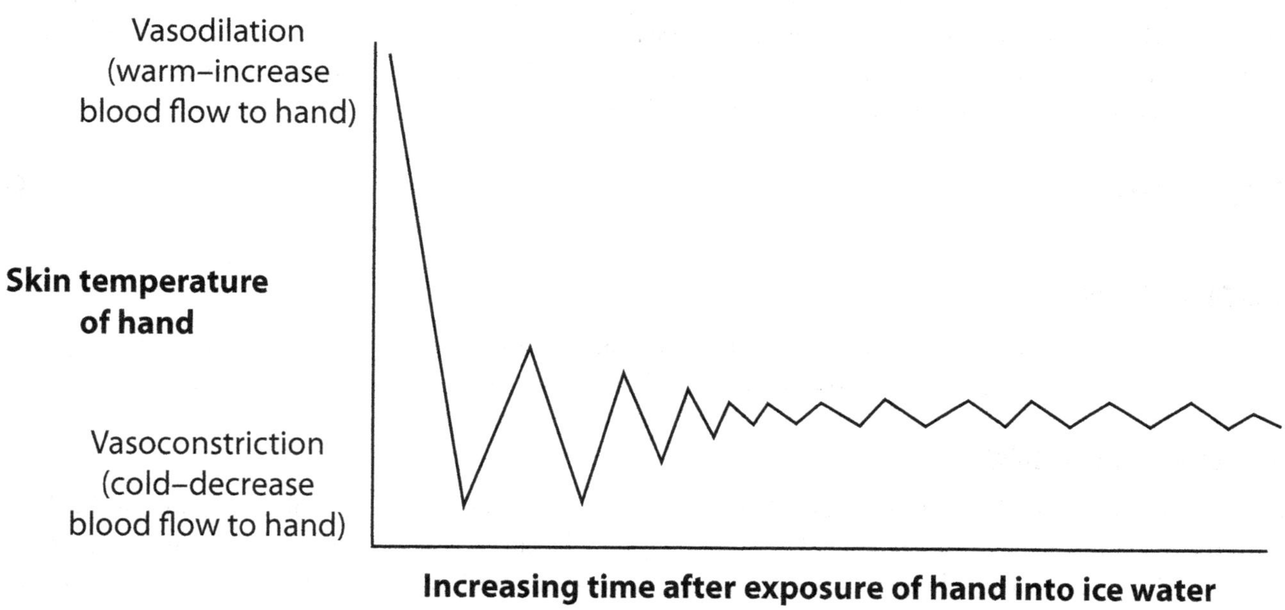

Figure 11.1. The cycle of vasoconstriction and vasodilation over time during cold stress. Modified from Frisancho 1993.

HEAT STRESS

When the body temperature rises rapidly above 98.6 degrees F, a condition called **hyperthermia** occurs. If this condition is not quickly reversed, then the individual may die. The goal is to increase heat loss. The physiological responses are sweating evaporation, which results in cooling of the body surface and vasodilation, which facilitates heat loss.

SKIN COLOR: PIGMENTATION AND DEPIGMENTATION

Melanocytes are specialized cells that produce the skin pigment **melanin**. All human beings have melanocytes (except albinos, as mentioned in the first paragraph of this chapter). But the difference in skin color, for example, between a person of European ancestry and a person of sub-Saharan African ancestry is the number and size of the melanin granules. In essence, the variation of skin color in human populations is related to the long period of adaptation of human populations in Africa, Europe, and Asia. In other words, the distribution of human skin color is an evolutionary process involving natural selection (Figure 11.2) [see Chapter 4].

What was the selective advantage for **pigmentation** (or dark skin) in regions of high solar radiation? We know that solar radiation is harmful in terms of sunburn, which can lead to serious infection, and skin cancer. But both of these effects make weak hypotheses for the evolution of dark skin simply because the effects of sunburn—particularly on the sweating mechanism—is still being studied and skin cancer usually occurs late in life (after one's reproductive years) therefore it has no impact on natural selection. Remember, natural selection is all about survival, reproduction, and passing your adaptive traits (genes) on to the next generation. Nonetheless, there has been a popular, long-standing hypothesis that linked the prevention of the overproduction of **vitamin D** with the selective advantage for dark skin (Loomis 1967). The necessary nutrient Vitamin D is synthesized naturally through the skin: skin + sunlight = vitamin D. It has been shown that vitamin D never reaches toxic levels in light-skinned individuals exposed to solar radiation over a long period of time (Holick et al. 1981).

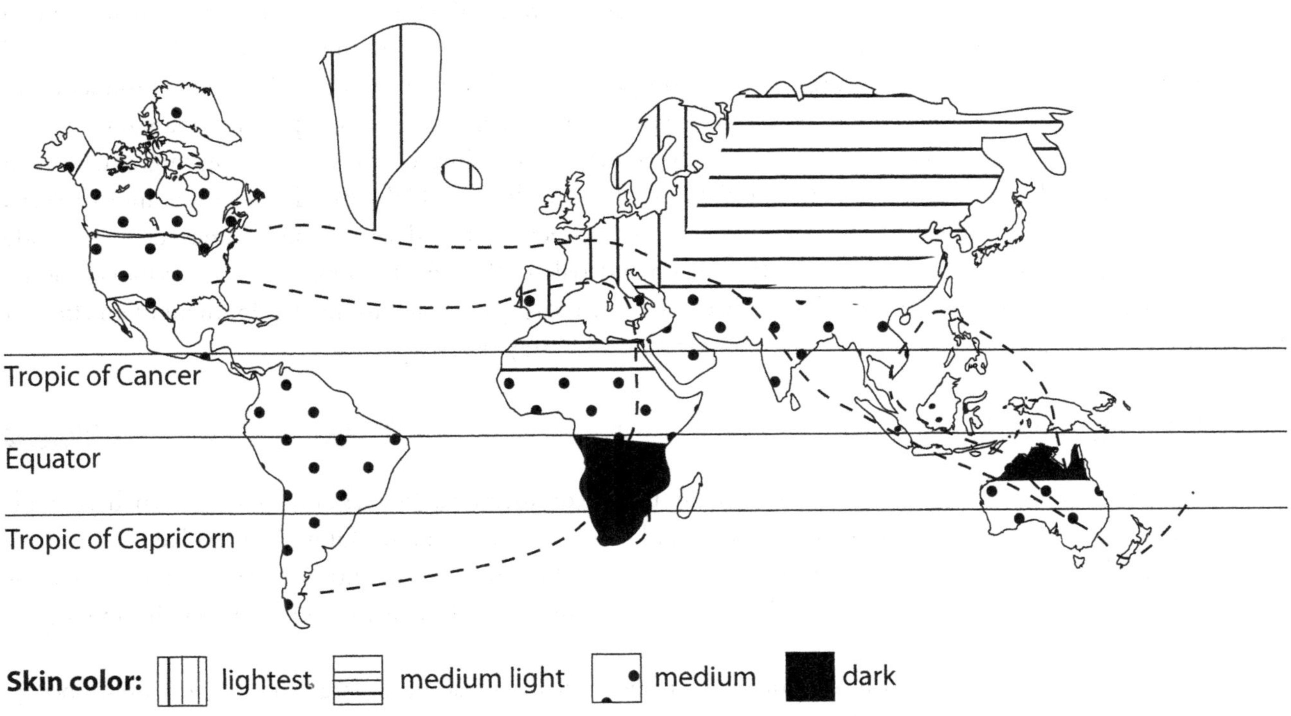

Figure 11.2. Distribution of human skin color in indigenous populations or populations whose ancestors have been in a region for one hundred thousand years or more. (Native American ancestry is in Asia [see chapter 10], but they have not been in the Americas long enough, particularly the South American groups, to become dark skinned).

Recently, researchers (Jablonski and Chaplin 2000) linked **folate** deficiency with the selective advantage for dark skin. Folate begins as folic acid, a necessary nutrient in the body. According to Jablonski and Chaplin, solar radiation destroys folate, and folate deficiency can lead to neural tube defects in the developing fetus, which is fatal. Also, folate deficiency can disrupt the production of sperm and cause male infertility. Dark skin or melanin protects folate from the harmful effects of solar radiation. Light skin (**depigmentation**) individuals who inhabit regions of high solar radiation for a long time, would suffer folate deficiency. Consequently, dark skin is favored by natural selection in areas of high solar radiation.

What happens to populations in areas of low solar radiation (cold climates)? We assume that the *Homo erectus* (see Chapter 9) populations that left Africa 2 million years ago were dark skinned. We know from the archaeology that *Homo erectus* populations were in the regions of low solar radiation, such as Europe and Northeast Asia. How did these populations survive? We know that dark skin blocks much of the harmful effects of solar radiation. We also know that vitamin D is necessary for normal bone development. Vitamin D deficiency causes diseases such as **rickets**, where bones fail to mineralize properly and bend (bowleg) due to the forces of body weight. Rickets is more critical in females because it impacts the pelvic bones and child birth. Again, the process of natural selection is affected. In essence, dark skin will be maladaptive because it blocks the little solar radiation that is available leading to vitamin D deficiency. Over time, a hundred thousand years or more, depigmentation would occur as an adaptation to regions of low solar radiation to facilitate the synthesis of vitamin D. Remember, variation is key to survival in changing environments (see Chapter 2). The human lineage has been successful for 2 million years because we have the genetic flexibility to adapt, i.e., skin color, in changing environments.

High Altitude Adaptation

There are many environmental stresses at high altitude (approximately 9,000 feet above sea level), such as cold; high solar radiation because of the closeness to the sun; and poor nutrition because not many fauna or flora can survive at high altitudes. However, the primary stress is **hypoxia,** where the body tissues are not receiving enough oxygen. Hypoxia results in headache, nausea, loss of appetite, fatigue, and breathlessness. We have already discussed 1) acclimatization or the physiological responses of most people, which is increase in red blood cell production in order to maintain sufficient oxygen levels in this low oxygen environment; and 2) developmental adaptation, where greater lung volume and the ability to use oxygen more efficiently develops over time in young migrants to high altitude environments. In contrast, human populations who have lived at high altitudes for thousands of years, such as the Peruvian highlanders or the Tibetan nomads, have larger chests and larger lung volumes relative to body size compared to populations at low altitude (Frisancho and Baker 1970; Beall 2006). But these populations are shorter compared with low-altitude populations because more energy is devoted to the development of an efficient respiratory system, leaving less energy for growth in stature and body size.

Biocultural Adaptation

The agricultural revolution (see Chapter 10) ushered in the age of infectious diseases and epidemics, such as smallpox, bubonic plague, tuberculosis, influenza, cholera, and malaria, to name a few. But we have evolved an immune system that enables us to adapt and survive amongst these lethal microbes. In addition, some random mutations are maintained in the population by natural selection to promote survival in this microbial environment, i.e. malaria-sickle cell (see Chapter 4).

Furthermore, the agricultural revolution also changed our diets leading to larger population sizes, obesity, and the emergence of non-infectious diseases.

NUTRITION AND DISEASE

As primates, humans are **omnivores**. This dietary flexibility enables us to adapt and survive in changing environments. Within this dietary flexibility, we must consume daily minimum amounts of carbohydrates, fats, proteins, vitamins, and minerals to maintain normal bodily functions. For example, lack of vitamin D can lead to rickets in children and **osteomalacia** in adults. And, as mentioned earlier, lack of folate leads to neural tube defects in the developing fetus.

There is a recommended daily allowance of calories and proteins for children and adults in the United States (see Figure 11.3). Logically, much energy and protein is required during growth and development and decline with age. More critically, pregnant and breast-feeding females need much more calories than average adult females. Obviously, there will be variation in energy and calorie requirements as the environment changes, i.e., high altitude environments or resource-poor environments (starvation).

In environments of starvation, the body concentrates the limited energy available on essential processes, such as maintaining the **basal metabolic rate** and sacrifices non-essential functions. For instance, the growth process is slowed down leading to small adult body size. In females, the age of the first menstrual cycle is delayed until late teens. The average age of the first menstrual cycle in well-nourished females is 11.7 years (U.S. females = 12.5 years) [Anderson et al. 2003]. Furthermore, there is mental retardation and susceptibility to infectious diseases due to a compromised immune system.

In contrast, overnutrition leads to an epidemic of non-infectious diseases. As stated earlier in this chapter, the agricultural revolution and the subsequent civilization it spawned had a negative effect on our diet. Today, fast food industries such as KFC, McDonalds, and Burger King, to name a few, coupled with globalism are changing the diets of traditional or "primitive" societies. In other words, these populations are becoming taller in stature. For instance, in Northeast Asia (China and Japan) we find a respectable number of the males 6 feet or taller. It is interesting that a 6 foot Chinese and Japanese in 1900 or 1950 was rare.

Continuing, the Polynesian island group Samoa provides an excellent example of the effect of modernization on human biology. Samoa is divided into Western Samoa and American Samoa. While Western Samoa has maintained much of its traditions, including diet, "American" Samoa has adopted the modern American culture and diet. A study conducted in 1985 found that males and females from American Samoa weighed more and had greater amounts of subcutaneous fat than males and females from Western Samoa (Bindon and Baker 1985; Relethford 2005). This new epidemic is found in many parts of the world and **globalism** is the vector.

In the United States, 50% of American adults and 20% of American children are overweight (Larsen 2008). And, this has led to health risks such as obesity, high blood pressure, and hypercholesterolemia. In short, the epidemic of non-infectious diseases such as type 2 diabetes, heart disease, kidney disease, stroke, and cancer, to name a few, is greater in the United States today than the prevalence of infectious diseases. In human evolution, there are costs and benefits. We benefited by having big brains to develop and enjoy the comforts of civilization. But the cost for these comforts is larger population sizes; changes in lifestyle, diet, physical activity, stress; and the emergence of infectious and non-infectious diseases. In essence, while the agricultural revolution was unavoidable—as far as nutrition and disease is concerned—it was probably the worst mistake in human history. So, throughout this text we see how cultural change significantly affects human biology and vice versa. This is the story of human origins.

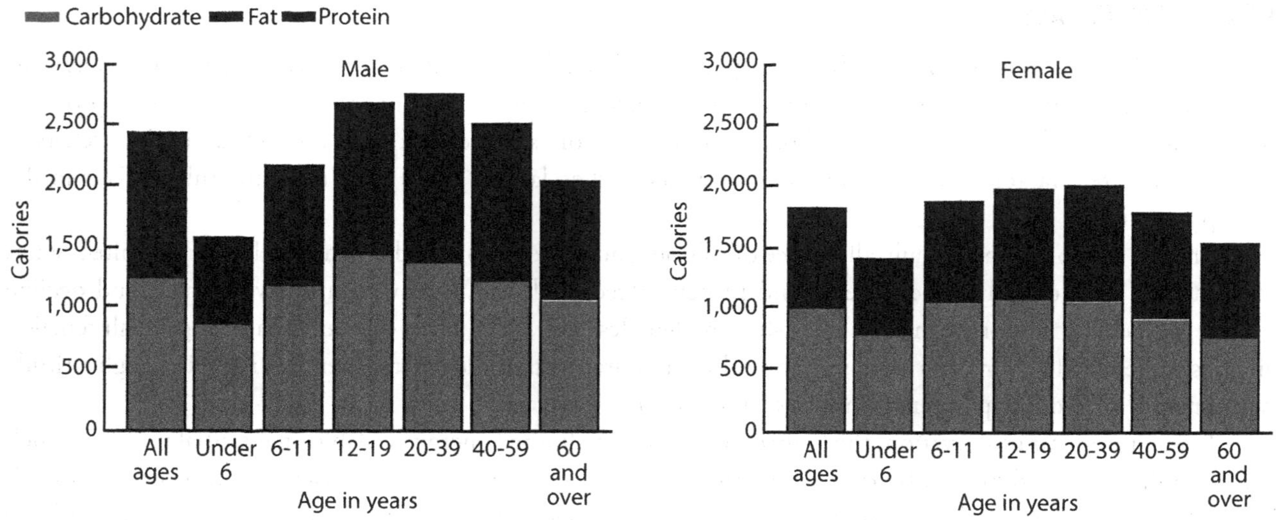

Figure 11.3 Recommended daily allowance of calories for children and adults in the United States.

Summary of Key Ideas

1. Skin color, hair form, and nose shape in present populations are traits inherited from our prehistoric ancestors hundreds of thousands of years ago. These traits are fixed—they will not change in a few generations.
2. Body size and shape can change in a few generations.
3. Genetic adaptations cannot be reversed. Developmental adaptations occur during growth and development, and cannot be reversed.
4. Physiological adaptations are short-term changes that are reversible.
5. Cultural adaptations are the tools/technology developed and used by people in a respective culture.
6. Vasoconstriction is the narrowing of blood vessels to conserve heat in the core of the body in response to cold temperatures. Vasodilation is the widening of blood vessels to facilitate heat loss in response to hot temperatures.
7. All human beings have melanocytes and the capacity to produce melanin (except for albinos). The differences in skin color between a sub-Saharan African and a European is in the number and size of the melanin granules.
8. Dark skin (pigmentation) evolved in regions of high solar radiation to protect folate. Excessive sunlight destroys folate, and folate deficiency is linked to problems in the developing fetus and male infertility. Both of these problems impact reproduction and, therefore, natural selection.
9. Light skin (depigmentation) evolved in regions of low solar radiation to facilitate the production of vitamin D. Vitamin D deficiency causes rickets, a bone deformity that occurs during growth and development. It is especially serious in females because a malformed pelvis impacts child birth and, therefore, natural selection.
10. People who live at high altitudes (genetic adaptation) or children born at low altitudes and raised at high altitudes (developmental adaptation) have efficient oxygen transport systems.
11. The agricultural revolution and subsequent civilizations, ushered in the age of infectious and non-infectious disease epidemics.

Key Terms

Basal Metabolic Rate
Climate Adaptation
Cultural Adaptation
Depigmentation
Developmental Adaptation
Folate
Genetic Adaptation
Globalism
High Altitude Adaptation
Human Adaptation
Hyperthermia
Hypothermia
Hypoxia
Melanin
Melanocytes
Omnivores
Osteomalacia
Physiological Adaptation
Pigmentation
Rickets
Vasoconstriction
Vasodilation
Vitamin D

Further Readings

Bogin, B. 1999. *Patterns of Human Growth.* 2nd edition. Cambridge, UK: Cambridge University Press.

Jablonski, N. G. 2006. *Skin: A Natural History.* Berkeley: University of California Press.

Moran, E. F. 1982. *Human Adaptability: An Introduction to Ecological Anthropology.* Boulder, CO: Westview Press.

Stinson, S., B. Bogin, R. Huss-Ashmore, and D. O'Rourke. 2000. *Human Biology: An Evolutionary and Biocultural Perspective.* New York: John Wiley & Sons.

Chapter 11: Modern Humans: Climate, High Altitude, and Cultural Adaptation

Study guide: By knowing the material below you will be well prepared for the exam. Don't forget to read the chapter and study your notes (this includes the chapter worksheets).

1. What is the initial physiological response to an elevated body temperature? What is a secondary response?
2. Discuss vasodilation and vasoconstriction as they relate to hot and cold environments
3. Be able define homeothermic
4. Be able to explain Bergmann and Allen's Rule for Adaptation
5. Bergmann's Rule states that heat-adapted mammals will have long limbs, which maximize the body's sufface area. T/F
6. Be able to define hypothermia
7. Be able to define basal metabolic rate
8. Be able to define melanocytes and melanin
9. Be able to discuss the vitamin D hypothesis for the evolution of light skin (depigmentation) in geographical regions (latitudes) where there is low UV radiation
10. Be able to discuss the weak and strong hypotheses for the evolution of dark skin (pigmentation) in geographical regions (latitudes) where there is high UV radiation
11. Be able to define hypoxia
12. Be able to define genetic adaptation; developmental adaptation; physiological adaptation; and cultural adaptation
13. Be able to discuss bio-cultural interaction as it relates to the problems of diet and the increase in non-infectious diseases.

NAME ______________________________ DATE ______________

Chapter 11: Modern Humans: Climate, High Altitude, and Cultural Adaptation

HOME WORK EXERCISES (10PTS.)

The goal of the home work exercises is assess your knowledge of the concepts and ideas presented in the chapter. Please answer the questions in each section as you go. When class lecture on this chapter is completed, carefully tear the home work exercises from the textbook and turn it in. **[Bonus pts. = 5]**

Section 11.1: Types of Adaptation

1. What is the difference between genetic adaptation and developmental adaptation? Use your own examples in your explanation.

2. What is the difference between acclimation and acclimatization?

3. Define and give four (4) examples of cultural adaptation.

Section 11.2: Climate Adaptation

1. Discuss how the human body physiologically responds to cold and heat stress.

2. Discuss the selective advantage for pigmentation (dark skin) in regions of high solar radiation.

3. Discuss the selective advantage for depigmentation (light skin) in regions of low solar radiation.

Section 11.3: High Altitude Adaptation

1. What are the environmental stresses at high altitude? Which one is the primary stress?

2. What type of adaptation do high altitude populations, such as the Peruvian highlanders and Tibetan nomads have? Describe their physique.

3. What type of adaptation do young migrants to high altitude, such as children born to low altitude parents have? Describe their physique.

4. If you went to visit friends at high altitude, how would your body respond to temporarily adapt you in this environmental stress?

Section 11.4: Biocultural Adaptation

1. Some historical scholars have argued that agriculture was the worst mistake in human history. Why?

2. From an evolutionary biology standpoint, are we still experiencing the effects of agriculture? Discuss. [Note: infectious and non-infectious diseases]

Glossary

ABO system: human blood group system consisting of the phenotypes A, B, AB, and O.

Acheulian tool technology: stone fashioned into a hand ax; biface; the tool technology of *Homo erectus*; a Middle Paleolithic tool, e.g., hand ax.

Adapis: one group of the "first primates" from the Eocene (56 to 34 mya) of Africa, Europe, Asia, and North America, technically known as euprimates, that might have been ancestral to lemurs and anthropoids.

Adaptation: traits or behaviors that allow organisms to survive and reproduce in a given environment; an optimal relationship between the organism and its environment.

Aegyptopithecus zeuxis: the well-known propliopithecid species from the Oligocene (34 to 24 mya) of North Africa and probably ancestral to catarrhines.

Agriculture: the practice of cultivating the soil; animal and plant domestication.

Allen rule: animals in hot climates have long limbs, which creates a larger surface area to volume in order to release heat. In contrast, animals in cold climates have short limbs, which creates a smaller surface area to volume in order to maintain heat.

Altruism: behaviors that benefit others at the expense of the actor.

Amino acid: organic molecules that are linked in specific ways to form proteins.

Amino acid dating: an absolute dating technique that measures the amount of change in amino acid structure; can be used to date bone.

Anagenesis: evolution within a single lineage; straight-line evolution; gradualism.

Angiosperm radiation hypothesis: forward-facing eyes, and grasping hands and feet evolved for finding and eating fruits made available with the radiation of flowering plants called angiosperms.

Anthropoidea: a suborder in animal taxonomy consisting of all monkeys, apes, and humans.

Anthropoids: (see Anthropoidea).

Antibodies: the primary defense in the immune system that attacks foreign antigens.

Antigens: foreign proteins, bacteria, blood cells that cause an immune reaction.

Arboreal: life in the trees.

Arboreal hypothesis: forward-facing eyes, and grasping hands and feet evolved for life in the trees.

Ardipithecus ramidus: one of the pre-australopiths dated 5.8 to 4.4 million years ago; it could be the missing link between chimpanzee and hominin lineages.

Auditory bulb: the portion of the temporal bone in the skull containing the ear bones.

Aurignacian: stone and bone tools associated with early modern humans; Upper Paleolithic tools, e.g., barbed harpoons, blades, projectile points, etc.

Australopithecines: the earliest hominins dated 4 to 1.5 million years ago; bipedal primates in the family Hominidae.

Australopithecus afarensis: hominin fossils dated 3.6 to 3.0 million years ago and found in East Africa; the most famous specimen is the 40% complete female skeleton found in Ethiopia nicknamed "Lucy."

Australopithecus africanus: hominin fossils dated approximately 2.5 million years ago and found in South Africa.

Australopithecus anamensis: a hominin fossil dated 4 million years ago and found in Kenya, East Africa.

Australopithecus garhi: a hominin fossil dated 2.5 million years ago and found in Ethiopia, East Africa; allegedly associated with the first tools.

Autosomal chromosomes: all of the chromosomes except the sex chromosomes; there are 22 pairs of autosomes.

Balanced polymorphism: selection for the heterozygotes against the homozygotes in discrete traits.

Balancing selection: selection against the homozygotes and for the heterozgotes in discrete traits.

Basal anthropoids: the root where all anthropoids originate; the common ancestor of all anthropoids.

Basal Metabolic rate: the energy required to maintain basic bodily functions at rest.

Bergmann rule: animals adapted to hot climates have small volume and large surface area to volume in order to release heat. In contrast, animals in cold climates have large volume and small surface area to volume in order to maintain heat.

Beringia: joins eastern Siberia to present day Alaska.

Biface: a stone tool fashioned so that it is pointed at one end; two sides of a core (raw material for stone tools) slanting inward to form a point.

Bilophodont molars: four raised cusps in line connected by a ridge of bone; dental trait in Old World monkeys.

Binomen: the two-name combination in scientific taxonomic nomenclature: *Homo* (*Genus*) *sapiens* (*species*).

Biocultural: the interaction between biology (genetic inheritance) and culture.

Biogeography: the correlation between the geographic distribution of animal and plant species and their biology (morphological characteristics).

Biostratigraphy: a relative dating technique where undated site A is assigned the same date as a well-known animal species (index fossil) found in many geographical regions (and at site A); technique used for sites where the composition of the rocks does not allow radiometric techniques.

Bipedalism: walking on two legs (humans are obligate bipeds).

***Biretia*:** a late-Eocene (37 mya) basal anthropoid, found in North Africa that might have been ancestral to anthropoids.

Blending inheritance: a 19th- and early 20th-century view that the "discrete particles" (or genes) blended (like mixing paint) in the offspring.

Bonobo: the pygmy chimpanzee; *Pan paniscus.*

***Branisella boliviana*:** the well-known parapithecid species from the Oligocene (34 to 24 mya) of South America, probably an early platyrrhine (ancient New World monkey).

Callitrichidae: a subfamily group in animal taxonomy consisting of a special group of New World monkeys; e.g., marmosets and tamarins.

Carbon-14 dating: a radiometric (absolute) dating technique where the decay rate of C-14 to N-14 is measured. The time it takes for half of C-14 to decay into N-14 is 5,730 years. A portion of the carbon-based item, i.e., bone, pottery, etc., must be destroyed in order to obtain the date.

***Carpolestes simpsoni*:** a plesiadapiform species or proprimate from the Paleocene (65 to 54 mya) of North America who had general primate traits such as grasping hands and feet, and nails instead of claws. This species has major implications for hypotheses of euprimate origins.

Catarrhines: (see Catarrhini).

Catarrhini: an infraorder in animal taxonomy consisting of Old World monkeys, apes, and humans.

Catastrophism: Georges Cuvier's hypothesis that sea levels rose due to global warming, causing a great flood that wiped out prehistoric life forms, such as the dinosaurs. After the flood, animals from other areas migrated into the vacant regions. In addition, God created new animals to "speed up" the migration process.

Cebidae: a family group in animal taxonomy consisting of a special group of New World monkeys; e.g., capuchin monkeys and squirrel monkeys.

Ceboidea: a superfamily group in animal taxonomy consisting of all New World monkeys.

Cell: contains several major structures, such as a nucleus, ribosomes, and mitochondria; the basic structure of tissues and organs.

Cenozoic: a geological era that dates from 65 million years ago to the present; the age of mammals.

Cercopithecidae: the subfamily group in animal taxonomy consisting of cercopithecoids and colobines.

Cercopithecoidea: a superfamily group in animal taxonomy consisting of all Old World monkeys.

Chemical dating: measuring the accumulation of chemicals in fossil materials. The more chemical absorbed by a fossil, the longer it has been in the ground and, therefore, the older it is, e.g., fluorine analysis.

Chimpanzee: the most intelligent of the great apes; genetically closest to humans; e.g., *Pan troglodytes* and *Pan paniscus.*

Chromosomes: strands of DNA containing thousands of genes.
Cladogenesis: the formation of two or more species; branching evolution.
Climate Adaptation: adaptation to temperature variation and high altitude.
Clovis Tradition: the earliest Native American culture of North America; the stone technology (projectile) is large, fluted, and bifacial.
Codon: three nucleotide bases (triplet) that code for a specific amino acid.
Colobinae: a subfamily group in animal taxonomy consisting of a special group of Old World monkeys; e.g., African colobines and Asian langurs.
Creationism: the belief that the universe was created by a Supreme Being, commonly referred to as "God" in human societies.
Creationists: people who believe that a Supreme Being created the universe.
Creation science: the scientific study of creationism (a false proposal).
Cro-Magnon: early modern humans dated 40,000 years ago and found in France; used as the model for modern human features.
Cross-fertilization: gametes from different individuals or different kinds (of plants) are fertilized.
Crossing over: in the first stage of meiosis, homologous chromosomes come together and exchange strands. Genes are recombined creating variation.
C_3 plants: Plants that take in carbon through C_3 photosynthesis, include trees, bushes, and shrubs associated with a relatively wet, wooded environment.
C_4 plants: Plants that take in carbon through C_4 photosynthesis, include open grassland, tropical savanna environment.
Cultural Adaptation: the tools/technology developed and used by people in a respective culture.
Culture: Learned behavior that is transmitted from individual to individual or generation to generation.
Dental anthropology: the study of the dental development, variation, and evolution in prehistoric and historic skeletal populations.
Depigmentation: light skin; present populations whose prehistoric ancestors evolved in regions far from the equator.
Developmental Adaptation: adaptations that are fixed during growth and development; a child born at low altitude but raised at high altitude will develop similar characteristics as the native highlander.
Diastema: a space next to the canine teeth; in animals with large canines the diastema provides space for the canine on the opposing jaw.
Diploid: in genetics it means 23 pairs of chromosomes, e.g., somatic cells.
Directional selection: a trait that is continually selected for because it is adaptive.
Diurnal: daytime activities, e.g., most monkeys are diurnal.
DNA: deoxyribonucleic acid; consists of a phosphate, sugar, and four nitrogenous bases. Codes for the biological characteristics of life forms.
Dominance: a ranking system in nonhuman primate societies.
Dominant: a trait that is expressed when the alleles are either in the heterozygote or homozygote forms.
Double helix: double-stranded DNA.
Dryopithecids: a group of mid-Miocene (13 mya) apes found in Africa, Europe, and Asia.
Endocast: impression of the inner surface of the skull
Enlightenment: this cultural period, also called the Age of Reason, began with full force in the 18th century (although its roots could be found as early as the 17th century) and ushered in the scientific revolution.
***Eosimias*:** a mid-Eocene (42 mya) basal anthropoid found in Asia that may be ancestral to anthropoids.
Estrus: a visible sign when a female animal is ready to mate; swelling in the female genital area, e.g., in dogs, chimpanzees, baboons, etc.
Euprimates: the first "true primates" emerging in the Eocene (56 to 34 mya).
Evolution: cumulative changes in morphological characteristics of a population over time (in few or many generations).

Fixity of species: a medieval/pre-Darwinian worldview of stasis in nature; there could not be change because God created a perfect world. Change (evolution) would mean that God was not perfect.

Folate: a necessary nutrient in the body that is important to the normal development of the fetus and normal male reproductive capabilities.

Foraminifera: ocean-dwelling microorganisms whose chemical compositions record temperature changes over time.

Forensic anthropology: the study of trauma on modern human skeletal remains and the estimation of race, sex, age, and stature (height) in suspicious deaths.

Founder effect: mutations that accumulate in a small population, separated from a larger population, because there is gene flow solely within the small group and none from outside of the group (any mutation within the group is passed from generation to generation).

FOXP2 gene: this gene has been identified in all mammals, and it may have mutated twice coinciding with the origins of speech and language in modern humans.

Frameshift mutation: the genetic message is rearranged and read incorrectly because of the substitution or deletion of nucleotides.

Galapagos Islands: these islands, off the northwest coast of Ecuador, South America, were made famous by Charles Darwin's study of finches.

Gemmules: Charles Darwin's designation for what we now call genes.

Gene: codes for amino acids and forms a polypeptide chain.

Gene flow: the exchange of genetic material between populations; migration; gene flow increases variation within populations and decreases variation between populations.

Genetic Adaptation: adaptations that are fixed/inherited and cannot be reversed; skin color, hair form, and nose shape are examples of genetic adaptations.

Genetic dating: DNA is used to show genetic relationships between different animal species, different human populations, or between humans and the apes.

Genetic drift: random change in gene frequencies from one generation to the next; random change in gene frequencies due to chance events; its greatest effect is in small populations; decreases variation within populations and increases variation between population.

Genotype: your genetic makeup.

Genus: the first name in the binomen; taxonomic nomenclature (e.g., *Homo*).

Gibbon: a lesser ape; genus: *Hylobates.*

Globalism: a world system linked economically and through mass media, and modern transportation systems.

Glutamic acid: an amino acid in the beta-hemoglobin chain.

God hypothesis: the proposition that an intelligent entity created the universe (this hypothesis cannot be tested or rejected; therefore, it is not a scientific hypothesis).

Gorilla: the largest of the great apes; genus: *Gorilla.*

Grand Designer: in creationism, it is a popular designation for God (referring to the fact that the universe or any complex structure within it could not have randomly evolved; there must be a Grand Designer).

Great Chain of Being *(Scalae Naturale)*: Aristotle's idea that all life was linked in a "great chain" with God on top, angels just below, and human races in ranked order below angels. Animals and inanimate objects are at the bottom of the hierarchy. This hierarchy was fixed.

Grooming: picking insects and dried skin from another primate's fur; integral to nonhuman primates' social life in reducing tension within the group.

Habitual bipedalism: consistently walking on two legs; obligate bipedalism; the locomotion of modern humans and their prehistoric ancestors.

Half-life: the time it takes for half of one element to decay into another element (or sometimes the same element).

Haplogroups: a set of related haplotypes that share similar mutations (see haplotypes).

Haploid: in genetics it means 23 chromosomes; sex cells; e.g., egg and sperm.

Haplorhini: includes all primates in the suborder Anthropoidea (all monkeys, apes, and humans); diurnal primates with reduction of the sense of smell (no wet nose).

Haplotypes: closely spaced loci (*sing.* Locus) on a single chromosome that are inherited together unchanged (no recombination); used as a molecular clock to determine evolutionary relationships.

Hardy-Weinberg equilibrium: a mathematical model that reflects the relationship between alleles and genotype frequencies; it is used to determine whether or not a population is undergoing evolutionary change.

Heliocentric solar system: a sun-centered solar system proposed by the 15th-century scientist Nicolaus Copernicus.

Heterodontic: different types of teeth; primates have incisors, canines, premolars, and molars.

Heterozygote: individuals with different alleles: dominant and recessive (Aa); a hybrid.

High Altitude Adaptation: adaptation to high altitude; large chest cavity and large lung volume are adaptations to a high altitude (low oxygen saturation).

Homogeneity: individuals that have similar phenotypic or genotypic characteristics.

Hominidae: the family group in animal taxonomy for living and dead human ancestors; australopithecines, *Homo habilis, Homo erectus,* Neanderthals (*Homo sapiens neanderthalensis*), modern *Homo sapiens.*

Hominoids: (see Hominoidea).

Hominin: bipedal prehistoric ancestors that date from 5 million years ago to the present.

Hominioidea: the superfamily in animal taxonomy consisting of apes and humans.

***Homo erectus*:** big-brained hominins dated 1.8 million to 300,000 years ago and found in Africa, Asia, Europe, and Indonesia; first world travelers who depended on culture, i.e, tools, clothing, fire, etc.

***Homo habilis*:** hominin fossils dated 2.0 to 1.8 million years ago from Kenya, East Africa; widely believed to be the makers of the first tools (Oldowan tools).

***Homo neanderthalensis*:** in hominin taxonomy, this binomen indicates that Neanderthals are a separate species, not in the lineage leading to modern humans.

***Homo sapiens neanderthalensis*:** in hominin taxonomy, this binomen indicates that Neanderthals are a subspecies or race.

Homozygote: individuals with the same alleles (AA).

Human biology: the study of the growth, development, variation, nutrition, illness, and disease in populations.

Human (*Homo sapiens*): Living and dead, habitual bipedal primates who depend on complex culture (speech and writing).

Human Adaptation: the varied ways humans cope in different environments; the favorable interaction between humans and their environments.

Human osteology: the study of growth and variation through the analysis of skeletal remains.

Human paleodemography: the statistical study of prehistoric population size, distribution, density, births, and deaths through the analysis of skeletal remains.

Human paleogenetics: the study of human origins/evolution and population relationships by obtaining bone samples for genetic analysis.

Human paleontology: the study of human origins/evolution through the analysis of fossilized primate skeletal remains showing evidence for bipedalism.

Human paleopathology: the study of disease and trauma in prehistoric populations through the analysis of skeletal and dental remains.

Hybrids: the offspring of two individuals of different species or two different races within the same species.

Hylobatidae: a family group in animal taxonomy consisting of the lesser apes.

Hyoid bone: a small bone in the throat that aids in the production of sounds necessary for speech.

Hyperthermia: when the body temperature rises rapidly above 98.6 degrees F.

Hypothermia: when the body temperature falls rapidly below 98.6 degrees F.

Hypothesis: a tentative statement used to explain a phenomenon in nature.

Hypoxia: the body tissues are not receiving enough oxygen; occurs at high altitudes.

Inheritance of acquired characteristics: in this hypothesis, an animal can generate a trait because of "need" during its lifetime and pass this acquired trait to the offspring. This hypothesis was proposed by Jean-Baptiste Lamarck.

Intelligent design (creationism): the idea that biological and non-biological structures, such as DNA, the eye, the solar system, etc., could not have appeared or evolve by chance or randomness. There must be a "Grand Designer"—a God.

Isotope: same number of protons and different number of neutrons.

Kin selection: behaviors that benefit the actor's relatives at the expense of the actor.

Knuckle walking: a quadrupedal locomotion where the weight of the arms rests on the knuckles of the hands; the locomotion of chimps and gorillas.

Lactose intolerance: some populations cannot tolerate milk because the enzyme lactase, which breaks down lactose (milk sugar), is inactive.

Law of independent assortment: traits sort independently of each other because the genes that code for them are on different chromosomes; the inheritance of one trait does not affect the inheritance of other traits.

Law of segregation: there are two alleles in every trait: one inherited from the mother and one inherited from the father.

Law of superposition: the strata (sedimentary layer) on the top is younger and the one below is older.

Levallois technique: preparing the core by knocking flakes off in several areas (of the core). Subsequently, flakes are knocked off and used for butchering animals; the technique of archaic *Homo sapiens.*

Lineage: ancestral-descendant linear relationships.

Locus: the position of a gene on a chromosome.

Malaria-sickle cell hypothesis: the sickle cell gene is in high frequencies in geographic regions where there is malaria because of natural selection. The malaria parasite cannot complete its life cycle in a sickle red blood cell; consequently, individuals with the trait (heterozygotes) survive and pass the gene to the next generation.

Medieval: the 10th to 15th centuries, often called the Dark Ages.

Megafauna: large game animals hunted by *Paleoindians*; woolly mammoth.

Melanin: a brown pigment that determines skin color; produced by melanocytes.

Melanocytes: specialized cells that produce the skin pigment melanin.

Meiosis: sex cell production.

Mendelian inheritance: the basis for genetics; simple inheritance involving the laws of segregation and of independent assortment.

Mendelian ratio: in the second generation of Gregor Mendel's cross-fertilization (of the pea plant), he got phenotypes of 3 to 1, e.g., 3 purple flowers to 1 white flower.

Mesozoic: this geological era dates from 245 to 65 million years ago; the age of reptiles.

Microevolution: evolution in a few generations so that the changes will be seen in future, living populations.

Miocene: time period dating from 25 to 5 million years ago during the Cenozoic era; the apes evolved in the Miocene.

Mitochondrial DNA (mtDNA): DNA located in the mitochondria of cells; mitochondrial DNA is passed down unchanged from female to female.

Mitosis: somatic cell production.

Modern synthesis: the synthesis of Mendelian genetics and Darwinian evolution.

Molecular anthropology: the study of the genetic distance between humans and the nonhuman primates, e.g., analyzing the genetic relationship between human populations.

Molecular clock: changes, such as mutations, that are so specific and regular that they can be used as a "clock" to date the last common ancestor of two populations or species, e.g., mitochondrial DNA.

Mousterian: stone tools associated with Neanderthals; Middle Paleolithic tools, e.g., Levallois flakes, scrapers, choppers, etc.

Mutation: a random change in the DNA; the only source of brand-new variation; increases variation between populations.

Nariokotome boy: the almost complete skeleton of a twelve-year-old *Homo erectus* boy dated 1.6 million years ago and found in Kenya, East Africa (West Turkana [WT] 15000).

Natural selection: in an environment of competition, those individual with adaptive traits will survive and reproduce and pass their traits to the offspring. There will be a proliferation of these adaptive traits in future generations. Those with nonadaptive traits will die.

Neanderthal: cold-adapted archaic *Homo sapiens* dated from 130,000 to 24,000 years ago and found in Europe and Asia; the first evidence of burials is found in Neanderthal culture.

New World monkey: monkeys indigenous to Mexico, Central and South America.

Nocturnal: nighttime activities, e.g., most prosimians are nocturnal.

Nonhoning: The posterior region of maxillary (upper) canines do not touch the anterior region of the first bicuspid (premolar); a characteristic of humans—small canines; the opposite of nonhuman primates and carnivores who have honing canines.

Nucleotide bases: the organic components of DNA; adenine-thymine, guanine-cytosine.

Numerical dating: (see Radiometric dating)

Old World monkeys: monkeys indigenous to Africa and Asia.

Oldowan complex: crude core tools dated approximately 2.5 million years ago; the first evidence of tools.

Oligocene: time period dating from 34 to 24 million years ago (mya) during the Cenozoic era; Old and New World monkeys evolved and diverged in the Oligocene.

Omomyids: one group of the "first primates" from the Eocene (56 to 34 mya) of Africa, Europe, Asia, and North America, technically known as euprimates, that might have been ancestral to tarsiers and anthropoids.

Omnivores: animals that eat meat and plant materials; many species in the primate order are Omnivores.

***On the Origin of Species*:** Charles Darwin's revolutionary monograph detailing his theory of evolution by natural selection, published in 1859.

Oreopithecus: a genus of dryopithecids found in Italy.

***Orrorin tugenesis*:** one of the pre-australopiths dated 6 million years ago; it could be the missing link between chimpanzee and hominin lineages.

Osteomalacia: a bone deformity in adults as a result of vitamin D deficiency.

Ouranopithecus: a genus of dryopithecids found in Greece.

Paleocene: time period dating from 65 to 54 million years ago during the Cenozoic era; Plesiadapiforms or "primate-like mammals" evolved in the Paleocene.

***Paleoindians*:** the earliest modern humans of the Americas; the first migration of Northeast Asian hunters-and-gatherers into North America 15,000 years ago.

Paleozoic: this geological era dates from 545 to 245 million years ago; the first reptiles.

Palmigrade: the palms of the hands are flat on the ground during quadrupedal locomotion like monkeys or human babies when they crawl.

Pangaea: a prehistoric supercontinent approximately 200 million years ago when all of the present continents were joined.

***Pan paniscus*:** the pygmy chimpanzee, commonly known as the bonobo; females are dominant in this species.

***Pan troglodytes*:** the larger, common chimpanzee; males are dominant in this species.

Phanerozoic eon: the geological time for the past 545 million years.

Parallel evolution: the independent evolution of similar adaptations in different species of animals; Old and New World monkeys evolved bilophodont molars, and front and back limbs that are the same size.

Parapithecids: a group of anthropoids from the Oligocene (34 to 24 mya) of North Africa, possibly ancestral to platyrrhines.

Phenotype: the physical expression of the gene; visible morphology.

Phylogenies: evolutionary histories of organisms.

Physical (biological) anthropology: the study of biological variation (which includes genetics) and evolution in living and dead human populations all over the world.

Physiological Adaptation: adaptations that are short-term and reversible; sweating or tanning are physiological adaptations.

Pigmentation: dark skin; present populations whose prehistoric ancestors evolved in regions on or close to the equator.

***Pithecanthropus erectus*:** the original genus name given by Eugene Dubois to fossil remains from Indonesia; *Homo erectus.*

Platyrrhines: (see Platyrrhini).

Platyrrhini: the infraorder in animal taxonomy consisting of all New World monkeys.

Pleiotrophy: a single gene coding for many traits or effects.

Pleistocene: the time period dating from 1.8 million to 10,000 years ago during the Cenozoic era; the genus *Homo* evolved during this period.

Plesiadapiforms: "primate-like mammals" or proprimates that evolved in the Paleocene (65 to 54 mya).

Pliocene: the time period dating from 5 to 1.8 million years ago during the Cenozoic era; the australopithecines evolved during this period.

Point mutation: a random substitution of a single nitrogenous base to another, which on occasion changes the amino acid.

Polygenic: several loci coding for a single trait or effect.

Polymorphism: a genetic trait with two or more alleles having frequencies greater than 1 percent.

Pongidae: the family group in animal taxonomy consisting of the great apes (orangutan, gorilla, and chimpanzee).

***Pongo pygmaeus*:** the orangutan.'

Post-orbital bar: a bony ring that separates the eye orbit from the skull.

Post-orbital closure: the eye orbit is surrounded by a "wall of bone" instead of a bony ring.

Potassium (40)-argon (40): a radiometric (absolute) dating technique that measures the decay rate of potassium (40)-argon (40); the time it takes for half of potassium (40) to decay into argon (40) is 1.3 billion years; the volcanic rock is dated, not the fossil.

Precambrian eon: the geological time dated from 4.6 billion years ago (the formation of the earth) to 545 million years ago (Phanerozoic eon).

Prehensile: grasping hands, feet, or tail; grasping tails are found in some New World monkeys and not in Old World monkeys.

Primates: an order in the class Mammalia defined by a suite of anatomical and behavioral traits including being large-brained, mostly tree-dwelling mammals that are highly social.

Primate social organization: the way nonhuman primates group themselves, e.g., single males; single females; male/female and offspring; single male/multifemale; multimale/multifemale; solitary.

Primatology: the scientific study of the living nonhuman primates.

Proconsulids: the early Miocene (22 mya) apes from Africa, possibly ancestral to later hominoids (apes and humans).

Propliopithecids: a group of anthropoids from the Oligocene (34 to 24 mya) of North Africa and probably ancestral to catarrhines (see *Aegyptopithecus zeuxis*).

***Propilopithecus*:** a propliopithecid genus from the Oligocene (34 to 24 mya) of North Africa.

Proprimates: the "primate-like mammals" (see Plesiadapiforms).

Prosimians: the most primitive primates; these primates have an enhanced sense of smell and nocturnal activity, e.g., lemurs, lorises, bush babies, etc.

Proteins: important complex molecules consisting of amino acids.

Proto-Cro-Magnons: archaic *Homo sapiens* dated 90,000 years ago from the Middle East (Western Asia) believed to be the descendants of the first moderns that migrated out of Africa 150,000 years ago, e.g., Qafzeh and Skhul.

Punnett square: a diagram used in Mendelian genetics that gives the proportion of different genotypes for the offspring after respective matings.

Quadramous: using all four limbs in climbing from tree to tree; orangutans bend the branches (or vault) as they use all four limbs to climb onto the next branch.

Quadrupedalism: walking on all four limbs.

Radiometric dating: a technique based on radioactive decay of unstable isotopes, e.g., carbon-14, potassium-40/argon-40, etc.

Recessive: a trait that is not expressed when the recessive gene is inherited from one parent but expressed when the recessive gene is inherited from both parents.

Recombination: homologous chromosomes come together during the first stage of meiosis and the genes on the chromosomes sort independently of each other, producing variation.

Regulatory genes: they turn other genes on and off; they regulate the expression of other genes.

Relative dating: a technique that does not give a number; it can only tell whether the fossil material is "older" or "younger."

Renaissance: the 15th to 18th centuries, often called the Age of Discovery.

Rhinarium: a moist pad on the nose; a wet nose; a trait that produces an enhanced sense of smell.

Rickets: a bone deformity in children as a result of vitamin D deficiency.

Robust australopithcines: hominins with very large faces, teeth, and jaws; their teeth reflect a diet of harsh, vegetative material.

Sacculated stomach: a special stomach that is divided in sections of sacs that enables some mammals to digest mature leaves with the aid of microorganisms.

***Sahelanthropus tchadensis*:** one of the pre-australopiths dated 7 to 6 million years ago; it could be the missing link between chimpanzee and hominin lineages.

Scientific method: involves setting up a hypothesis and testing this hypothesis through data collection, experimentation, observation, and measurement. The hypothesis must be testable and subject to rejection.

Sedentism: nonmigratory; settled in one place, e.g., agricultural populations.

Sexual dichromatism: differences in color of hair and skin between males and females of mammalian species, e.g., male and female gibbons have different-colored fur.

Sexual dimorphism: differences in the size of certain morphological features, e.g., body or canine teeth, etc., between males and females of mammalian species, e.g., male baboons, orangutan, and gorillas are twice as large as their female counterparts.

Sickle cell: a point mutation in the hemoglobin that results in a sickle red blood cell.

Sickle cell anemia: a fatal genetic disease where the affected individual has inherited the sickle gene from both parents (heterozygous dominant or carriers); this individual is homozygous recessive.

Sickle cell gene: a point mutation changing the normal amino acid glutamic acid (CTC) to valine (CAC).

***Sivapithecus*:** a genus of dryopithecids found in Asia and probably ancestral to orangutans.

Speciation: the formation of new species.

Species: individuals that can interbreed and produce a fertile offspring; constant gene flow is critical in animal species because it prevents the formation of new species.

Stabilizing selection: selection against the extremes and for the average in continuous traits.

Stable carbon isotopes: the isotopes are not unstable, e.g., carbon-12 and carbon-13, etc.

Stable oxygen isotopes: the isotopes are not unstable, e.g., oxygen-16 and oxygen-18, etc.

Stasis: fixed or unchanging (as in the pre-Darwinian worldview of nature).

Stereoscopic binocular vision: overlapping fields of vision that give both depth perception and distance.

Strata: the sedimentary layers; the layers on top are younger than the layers below.

Stratigraphy: a relative dating technique based on the law of superposition: the layers above are younger than the layers below.

Strepsirhini: includes all primates in the suborder Prosimii (lemurs, lorises, bush babies); these primates are generally nocturnal and have an enhanced sense of smell, e.g., prosimians have a rhinarium.

Synonymous mutation: changes the codon but does not result in a change of the amino acid; a neutral mutation.

Taxonomy: classifying plants and animals into categories according to similarities and differences. Carolus Linnaeus is the father of taxonomy.

Terrestrial: life on the ground.

Theory: a hypothesis that has been tested several times independently and not rejected.

Thermoluminescence: an absolute dating technique based on trapped electrons; the fossil material, i.e., burned flint, must be collected and taken back to the lab where it will be reheated to a specific temperature. The luminescence represents the trapped electrons.

Transcription: DNA makes a copy of itself.

Translation: the genetic message is translated into amino acids and proteins; protein synthesis.

True-breeding: pure; homozygous.

Uniformitarianism: natural forces, such as wind erosion, earthquakes, volcanic eruptions, etc., that presently affect the earth are the same forces that have occurred in the past. They are uniform and very slow in changing the landscape. This theory was proposed by Charles Lyell and James Hutton.

Use and disuse: traits that are frequently used become stronger and traits that are not used gradually become weaker and disappear. This hypothesis was proposed by Jean-Baptiste Lamarck.

Vasoconstriction: narrowing of the blood vessels reducing blood flow and preventing heat loss from the core of the body.

Vasodilation: widening of the blood vessels increasing blood flow and facilitating heat loss from the core of the body.

Victoriapithecids: a group of anthropoids from the late Oligocene/early Miocene (23 mya) of Africa, possibly ancestral to Old World monkeys.

Visual predation hypothesis: forward-facing eyes, and grasping hands and feet evolved for hunting insects in an arboreal environment.

Vitamin D: an important nutrient that is critical to normal bone development; vitamin D deficiency causes rickets in children and osteomalacia in adults.

Y-5 molar pattern: on the occlusal surface of the molar a "Y" groove is formed between the five cusps; this cusp pattern is found in apes and humans.

References

Anderson, S. E., G. E. Dallal, A. Must. 2003. Relative weight and race influence average age at menarche: Results from two nationally representative surveys of U.S. girls studied over 25 years apart. *Pediatrics* 111: 844-850.

Baker Publishing Group. 1995. *Holy Bible: King James version.* Standard text edition. New York: Cambridge University Press.

Beall, C. M. 2006. Andean, Tibetan, and Ethiopian patterns of adaptation to high altitude hypoxia. *Integrative and Comparative Biology* 46: 18-24.

Begun, D., and A. Walker. 1993. The endocast. In *The Nariokotome Homo erectus skeleton,* ed. A. Walker and R. Leakey, 326–58. Cambridge, MA.: Harvard University Press.

Benefit, B. R. and M. McCrossin. 1991. Ancestral facing morphology of Old World higher primates. *Proceedings of the National Academy of Sciences USA* 88: 5267-5271.

Berger, T. D., and E. Trinkaus. 1995. Patterns of trauma among the Neandertals. *Journal of Archaeological Science* 22: 841–52.

Bindon, J. R. and P. T. Baker. 1985. Modernization, migration, and obesity among Samoan adults. *Annuals of Human Biology* 12: 67-76.

Biro, D., and T. Matsuzawa. 2001. Chimpanzee numerical competence: Cardinal and ordinal skills. In *Primate origins of human cognition and behavior,* ed. T. Matsuzawa, 199–225. New York: Springer.

Boaz, N. T., and A. J. Almquist. 2002. *Biological anthropology: A synthetic approach.* Upper Saddle River, NJ: Prentice Hall.

Boyd, R., and J. B. Silk. 2006. *How humans evolved,* 4th ed. New York: W. W. Norton.

Brace, C. L. 1996. A four-letter word called 'race.' In *Race and other misadventures: Essays in honor of Ashley Montagu in his ninetieth year,* ed. L. T. Reynolds & L. Lieberman, 106–41. New York: General Hall Publishers.

Brace, C. L. 2005. *Race is a four-letter word: The genesis of the concept.* New York: Oxford University Press.

Brace, C. L., and A. Montagu. 1977. *Human evolution: An introduction to biological anthropology,* 2nd ed. New York: Macmillan.

Campbell, B. G., J. D. Loy, and K. Cruz-Uribe. 2006. *Humankind emerging,* 9th ed. New York: Allyn & Bacon.

Cann, R. L., M. Stoneking, and A. C. Wilson. 1987. Mitochondrial DNA and human evolution. *Nature* 325: 31–36.

Carlson, K.J., D. Stout, T. Jashashvili, D.J. de Ruiter, P. Tafforeau, K. Carlson, L.R. and Berger. 2011. The endocast of MH1 *Australopithecus sediba.* Science 333:1402-1407

Cartmill, M. 1974. Rethinking primate origins. *Science* 184: 436-443.

Cavalli-Sforza, L. L., P. Menozzi, and A. Piazza. 1994. *The History and Geography of Human Genes.* Princeton: Princeton University Press.

Clark, J.D., J. de Heinzelin, K.D. Schick, W.K. Hart, T.D. White, G. WoldeGabriel, R.C. Walter, G. Suma, B. Asfaw, E. Vrba, and Y.H. Selassie 1994 African *Homo erectus:* Old radiometric ages and young Oldowan assemblages in the Middle Awash Valley, Ethiopia. Science 264: 1907-1910

Cowen, R. 1995. *History of life,* 2nd ed. Boston: Blackwell.

Dart, R. 1925. *Australopithecus africanus*: The man-ape of South Africa. *Nature* 115: 195–99.

Darwin, C. 1962. *The voyage of the* Beagle. Garden City, NY: Doubleday. (First published in 1839.)

Darwin, C. 1981. *The descent of man, and selection in relation to sex.* Princeton: Princeton University Press. (First published in 1871.)

Dawkins, R. 2006. *The God delusion.* New York: Houghton Mifflin.

Day, M. 1965. *Guide to fossil man.* Cleveland: World Publishing.

De Heinzelin, J., J.D. Clark, K.D. Schick, and W.H. Gilbert. 2000. The Achuelian and the Plio-Pleistocene deposits of the Middle Awash Valley, Ethiopia. Musée Royal de L' Afrique Centrale, Annales Sciences Geologiques 104.

De Heinzelin, J., J.D. Clark, T.D. White, E. Hart, P. Renne, G. WoldeGabriel, Y. Beyene, and E. Vrba.. 1999. Environment and behavior of 2.5 million-year-old Bouri hominids. *Science* 284: 625-629.

de Waal, F. B. 1995. Bonobo sex and society. *Scientific American* 272: 82–88.

de Waal, F. B. 1999. Cultural primatology comes of age. *Nature* 399: 635–36.

Elmore, B. 2007. Defining faith: Landmark national research by Baylor faculty sheds new light on America's view of god, religion, morality, politics. *Baylor Magazine* 5: 24–29.

Fleagle, J. 1999. *Primate adaptation and evolution*, 2nd ed. San Diego: Academic Press.

Fox, E. A., and I. bin'Muhammad. 2002. New tool use by wild Sumatran orangutans (*Pongo pygmaeus abelli*). *American Journal of Physical Anthropology* 119: 186–88.

Frisancho, A. R. 1993. *Human Adaptation and Accommodation*. Ann Arbor: University of Michigan Press.

Frisancho, A. R. and P. T. Baker. 1970. Altitude and growth: A study of the patterns of physical growth of a high altitude Peruvian Quechua population. *American Journal of Physical Anthropology* 32: 279-292.

Gardner, R. A., and Gardner, B. T. 1969. Teaching sign language to a chimpanzee. *Science* 165: 664–72.

Gibbons, A. 2011. Skeletons present an exquisite paleo-puzzle. *Science* 333: 1370-1372.

Gibbons, A. 2011 Who were the Denisovans? *Science* 333: 1084-1087.

Gingerich, P. D. 1976. Paleontology and phylogeny: Patterns of evolution at the species level in early tertiary mammals. *American Journal of Science* 276: 1-28.

Gingerich, P. D. 1986. Plesiadapis and the delineation of the order primates. In *Major Topics in Primate and Human Evolution,* ed. B. Wood, L. Martin, and P. Andrews, pp. 32-46. Cambridge, UK: Cambridge University Press.

Gingerich, P. D. 1990. Mammalian order proprimates. *Journal of Human Evolution* 19: 821-822.

Goodall, J. 1986. *The chimpanzees of Gombe: Patterns of behavior*. Cambridge, MA: Harvard University Press.

Gould, S. J. 1987. *Time's arrow, time's cycle*. Cambridge, MA: Harvard University Press.

Gould, S. J. 1994. The geometer of race. *Discover* 15: 65–69.

Green, R.E., J. Krause, S.E. Ptak, A.W. Briggs, M.T. Ronan, J.F. Simons, L. Du, M. Egholm, J.M. Rothberg, M. Paunovic, and S. Pääbo. 2006. Analysis of one million base pairs of Neanderthal DNA. *Nature* 444: 330-336.

Hall, Stephen. 2008. Last of the Neanderthals: Why did our Ice Age rivals vanish? *National Geographic* 214: 37-59.

Henry, A.G., A.S. Brooks, and D.R. Piperno. 2011. Microfossils in calculus demonstrate consumption of plants and cooked foods in Neanderthal diets (Shanidar II, Iraq; Spy I and II, Belgium). Proceedings of the National Academy of Sciences USA 108: 486-491.

Holick, M. F., J. A. MacLaughlin, and S. H. Doppett. 1981. Regulation of cutaneous previtamin D_3 photosynthesis in man: Skin pigmentation is not an essential regulator. *Science* 211: 590-593.

Houle, A. 1999. The origin of platyrrhines: An evaluation of the Antarctic scenario and the floating island model. *American Journal of Physical Anthropology* 109: 541-559.

Hoyle, F., and C. Wickramasinghe. 1981. *Evolution from space: A theory of cosmic creationism.* New York: Simon & Schuster.

Isbell, L., and T. Young. 1996. The evolution of bipedalism in hominids and reduced group size in chimpanzees: Alternate responses to decreasing resource availability. *Journal of Human Evolution* 30: 389–97.

Jablonski, N. G. and G. Chaplin. 2000. The evolution of human skin coloration. *Journal of Human Evolution* 39: 57-106.

Jurmain, R., and H. Nelson. 1994. *Introduction to physical anthropology,* 6th ed. New York: West.

Karafet, T. M., S. L. Zegura, O. Posukh, A. Bergen, J. Long, D. Goldman, W. Klitz, S. Harihara, P. de Knijff, V. Wiebe, R. C. Griffiths, A. R. Templeton, and M. F. Hammer. 1999. Ancestral Asian source(s) of New World Y-chromosome founder haplotypes. *American Journal of Human Genetics* 64: 817-831.

Kibii, J.M., S.E. Churchill, P. Schmid, K.J. Carlson, N.D. Reed, D.J. de Ruiter, and L.R. Berger. 2011. A partial pelvis of *Australopithecus sediba. Science* 333: 1407-1411.

Kivell, T.L., J.M. Kibii, S.E. Churchill, P. Schmid, and L.R. Berger. 2011. *Australopithecus sediba* hand demonstrates mosaic evolution of locomotor and manipulative abilities. *Science* 333: 1411-1417.

Krause, J., Q. Fu, J.M. Good, B. Viola, M.V. Shunkov, A.P. Derevianko, and S. Pääbo 2010. The complete mitochondrial DNA genome of an unknown hominin from southern Siberia. *Nature* 464: 894-897.

Krings, M., A. Stone, R. W. Schmitz, H. Krainitzki, M. Stoneking, and S. Pääbo. 1997. Neandertal DNA sequences and the origin of modern humans. *Cell* 90: 19–30.

Larsen, C. S. 2008. *Our origins: Discovering physical anthropology.* New York: W. W. Norton.

Linden, E. 1981. *Apes, men, and language.* Middlesex, England: Penguin Books.

Lindroth, S. 1973. Linnaeus (or von Linne), Carl. P. In *Dictionary of scientific biography*, vol. 8, ed. C. C. Gillispie, 380. New York: Charles Scribner's Sons.

Loomis, W. F. 1967. Skin-pigment regulation of vitamin-D biosynthesis in man. *Science* 157: 501-506.

Lovejoy, A. O. 1961. *The great chain of being: A study in the history of an idea.* The William James Lectures 1933. Cambridge, MA: Harvard University Press.

Lovejoy, C. O. 1981. The origin of man. *Science* 211: 341–50.

McGrew, W. C. 1992. *Chimpanzee material culture.* Cambridge, MA: Cambridge University Press.

McGrew, W. C. 1998. Culture in nonhuman primates? *Annual Review of Anthropology* 27: 301–28.

Mercader, J., M. Panger, and C. Boesch. 2002. Excavation of a chimpanzee stone tool site in the African rainforest. *Science* 296: 1452–55.

Molnar, S. 1998. *Human Variation:* Races, Types, and Ethnic Groups, 4th ed. Englewood Cliffs, NJ: Prentice-Hall

Nystrom, P., and P. Ashmore. 2008. *The life of primates.* Upper Saddle River, NJ: Pearson Education.

Patterson, F. 1978. Linguistic capabilities of a lowland gorilla. In *Sign language and language acquisition in man and* ape, ed. F. Peng, 161–201. Boulder, CO: Westview Press.

Patterson, F., and E. Linden. 1981. *The education of Koko.* New York: Holt, Rinehart and Winston.

Pickering, R., P.H. Dirks, Z. Jinnah, D.J. de Ruiter, S.E. Churchill, A.I. Herries, J.D. Woodhead, J.C. Hellstrom, and L.R. Berger. 2011. *Australopithecus sediba* at Ma and implications for the origins of the genus *Homo. Science* 333: 1421-1423.

Pruetz, J. D., and P. Bertolani. 2007. Savannah chimpanzees, *Pan troglodytes verus*, hunt with tools. *Current Biology* 17: 412–17.

Raven, P. H., and G. B. Johnson. 2002. *Biology*, 6th ed. New York: McGraw-Hill.

Reich, D., R.E. Green, M. Kircher, J. Krause, N. Patterson, E.Y. Durand, B. Viola, A.W. Briggs et al. 2010. Genetic history of an archaic hominin group from Denisova Cave in Siberia. *Nature* 4688: 1053-1060.

Relethford, J. H. 2005. *The human species: An introduction to biological anthropology*, 6th ed. New York: McGraw-Hill.

Rodman, P., and H. McHenry. 1980. Bioenergetics and the origin of hominid bipedalism. *American Journal of Physical Anthropology* 52: 103–6.

Rumbaugh, D. M. 1977. *Language learning by a chimpanzee: The Lana project.* New York: Academic Press.

Sakolsky, J. 2005. *Copernicus and modern astronomy.* New York: Rosen Publishing Group.

Savage-Rumbaugh, S., and R. Lewin. 1994. *Kanzi: The ape at the brink of the human mind.* New York: John Wiley & Sons.

Savage-Rumbaugh, S., S. G. Shanker, and T. J. Taylor. 1998. *Apes, language, and the human mind.* New York: Oxford University Press.

Schurr, T. G. 2000. Mitochondrial DNA and the peopling of the New World. *American Scientist* 88: 246-253.

Smith, B. D. 1995. *The Emergence of Agriculture.* New York: Scientific American Library, W. H. Freeman.

Snowden, C. T. 1990. Language capacities of nonhuman animals. *Yearbook of Physical Anthropology* 33: 215–43.

Stringer, C. B., and P. Andrews. 1988. Genetic and fossil evidence for the origin of modern humans. *Science* 239: 1263–68.

Stringer, C. B., and R. McKie. 1996. *African exodus: The origins of modern humanity.* New York: Henry Holt.

Sussman, R.W. 1991. Primate origins and the evolution of angiosperms. *American Journal of Primatology* 23: 209-223.

Szalay, F.S., A. L. Rosenberger, and M. DaGosto. 1987. Diagnosis and differentiation of the order primates. *Yearbook of Physical Anthropology* 30: 75-105.

Terrace, H. S. 1987. *Nim.* New York: Columbia University Press.

Tyson, N. 1999. Holy wars. *Natural History* 108: 80–82.

Tyson, N. 2005. The perimeter of ignorance: A boundary where scientists face a choice, invoke a deity or continue the quest for knowledge. *Natural History* 114: 28–34.

Ungar, P.S. and M. Sponheimer. 2011. The diets of early hominins. *Science* 334: 190-193.

U.S. Department of Health and Human Services (2000). Intake of calories and selected nutrients for the U.S. population 1999-2000. Retrieved February 1, 2011 from http://www.fda.gov.

Van Schaik, C. P., M. Ancrenaz, G. Borgen, B. Galdikas, C. D. Knott, I. Singleton, A. Suzuki, S. S. Utami, and M. Merrill. 2003. Orangutan cultures and the evolution of material culture. *Science* 299: 102–5.

Van Schaik, C. P., and C. D. Knott. 2001. Geographic variation in tool use on *Neesia* fruits in orangutans. *American Journal of Physical Anthropology* 114: 331–42.

Wheeler, P. 1993. Human ancestors walked tall, stayed cool. *Natural History* 102: 65–67.

White, T., B. Asfaw, Y. Beyene, Y. Haile-Selassie, C.O. Lovejoy, G. Suwa, and G. WoldeGabriel. 2009. *Ardipithecus ramidus* and the paleobiology of early hominids. *Science* 326: 75-86.

Wolpoff, M. H., A. G. Thorne, F. H. Smith, D. W. Frayer, and G. G. Pope. 1994. Multiregional evolution: A worldwide source for modern human populations. In *Origins of anatomically modern humans,* ed. M. H. Nitecki and D. V. Nitecki, 175–200. New York: Plenum Press.

Zipfel, B., J.M. DeSilva, R.S. Kidd, K.J. Carlson, S.E. Churchill, and L.R. Berger. 2011. The foot and ankle of *Australopithecus sediba.* Science 333: 1417-1420.

Index

A

ABO blood group system, 30–31, 32
Acclimation, 147–148
Acheulian tool technology, 123
Adapids, 54–55
Adaptation, 11
  biocultural, 150–152
  climate, 147, 148–150
  cultural, 148
  developmental, 147
  genetic, 147
  high altitude, 147, 150
  human adaptation, 147
  physiological, 147–148
Adaptation, rules of, 134
Aegyptopithecus zeuxis, 56
Age of Discovery, 10
"Age of Mammals," 54
Age of Reason, 10–11
"Age of Reptiles," 89
Agriculture, early, 137
Alleles, 24
Allen rule of adaptation, 134
Almquist, A. J., 13, 15
Amino acid dating, 92
Amino acids, 43
Anagenesis, 131
Anderson, S. E., 151
Andrews, P., 131
Angiosperm radiation hypothesis, 53, 54
Animal domestication, 137
Anthropoidea, 70–80
Anthropoids, 55–59
Antibodies, 30
Antigens, 30
Apes, 57
Arboreal environment, 67
Arboreal hypothesis, 53–54
Ardipithecus ramidus, 103, 104
Argon-40, 92
Ashmore, P., 55, 56, 57, 67
Assimilation hypothesis, 133
Atapuerca (Spain), 124
Auditory bulb, 54
Aurignacian complex, 137
Australopithecines, 105–109
Australopithecus aethiopicus, 108
Australopithecus afarensis, 105
Australopithecus africanus, 106
Australopithecus anamensis, 105
Australopithecus boisei, 108
Australopithecus garhi, 109
Australopithecus robustus, 108
Australopithecus sediba, 106
Australopiths and early *Homo,* 101–119
  australopithecines, 105–109
  diets of early hominims, 111–114
  early *Homo,* 110–111
  evidence of tool use in Australopiths, 109–110
  origin of bipedalism, 101–102
  pre-australopithecines, 102–105
Autosomal chromosomes, 27

B

Baboons, 69
Bacon, F., 10
Baker, P. T., 150, 151
Balanced polymorphism, 45
Balancing selection, 44–45
Barbed harpoons, 137
Basal anthropoids, 55
Basal metabolic rate, 151
Beall, C. M., 150
Begun, D., 123
Benefit, B. R., 57
Berger, Lee, 106, 107, 108
Berger, T., 135
Bergmann rule of adaptation, 134
Beringia, 137
Bertolani, P., 2
Biface tools, 123
Bilophodont teeth, 72
Bindon, J. R., 151
Bin'Muhammad, I., 2
Binomen, defined, 70
Biocultural, defined, 1
Biocultural adaptation, 150–152
Biogeography, 67
Biological anthropology. *see* physical (biological) anthropology
Biostratigraphy, 91
Bipedalism, ix, 67, 101–102
Biretia, 44
Biro, D., 3
Blades, 137
Blending inheritance, 23
Blood polymorphisms, 30–31
Boaz, N. T., 13, 15
Bodo (Ethiopia), 124
Body temperature, bipedalism and, 102
Bone tools, 137
Bonobos, 78
Boule, Marcellin, 133
Boxgrove (England), 124
Boyd, R., 23, 25
Brace, C. L., 9–11
Branisella boliviana, 56
Buffon, Comte de, 12–13
Burins, 137

C

Campbell, B. G., 11, 12–13, 15, 23, 25, 102, 122, 134
Cann, R. L., 131
Carbon-14 dating method, 91–92
Carbon isotopes, 113–114
Carpolestes simpsoni, 54
Cartmill, M., 54
Carved figurines, 137
Catarrhine evolution, 57–59
Catarrhini, 70
Catastrophism, 13
Catayrrhines, 55
Cavalli-Sforza, L. L., 138
Cave paintings, 137
Caves/temporary shelters, 123
Cebidae, 71
Ceboidea, 71
Cells, 26
Cenozoic era, 54, 89, 90
Cercopithecidae, 57
Chaplin, G., 150
Chemical dating, 91
Chimpanzee 1 (common chimpanzee), 67, 76, 77
Chimpanzee 2 (bonobos), 78
Chromosomes, 26, 27
Cladogenesis, 133
Climate adaptation, 147, 148–150
Clovis Tradition, 138
Codons, 43
Colobinae, 57, 71
Copernicus, N., 10
Correns, C. E., 26
Cowen, R., 89
"Creation science," 3
Crick, F., 26
Cro-Magnons, 136–137
Cross-fertilization, 23–24
Crossing over, 28
Cultural adaptation, 148
Culture, 1
Cuvier, G., 13, 14

D

Dart, R., 101
Darwin, C.
  On the Origin of Species, 11–12, 15
  pre-Darwinian thought, 9–11
  theory of natural selection, 14–16, 23, 43
  tool use/hunting hypothesis of, 101
Dating methods, 89–99
  Cenozoic era, 90
  geological time periods, defined, 89, 91

Mesozoic era, 89–90
numerical dating (nonradiometric), 92
numerical dating (radiometric), 91–92
Paleozoic era, 89
reconstruction of ancient environments, 92–94
relative dating, 91
Dawkins, R., 3
Day, M., 102, 134
De Laplace, P. S., 2
Denisovans, 135–136
Dental anthropology, 2
Dental microwear analysis, 111–113
Depigmentation, 150
Derevianko, A., 135
Developmental adaptation, 147
De Vries, H., 25
De Waal, F. B., 2
Diastema, 69
Diet
dietary flexibility, 69
of early hominins, 111–114
Homo erectus, 123
Diploid, 27
Directional selection, 44
Diurnal activities, 68
DNA
discovery of, 14
mitochondrial DNA (mtDNA), 132–133
molecular genetics, 26
Neanderthal, 135
Dominance, 24, 77
Double helix, 26
Dryopithecids, 57–58
Dubois, E., 121

E

Early *Homo (Homo habilis)*
diets of early hominims, 111–114
generally, 110–111
Homo erectus and, 122
Elmore, B., 3
Energy efficiency, bipedalism hypothesis and, 101–102
Enhanced sense of touch, 67
Enhanced sense of vision, 67
Eocene epoch, 54
Eosimias, 44
Essay on the Principle of Population, An (Malthus), 14
Estrus cycle, 76
Euprimates, 54–55
Eve hypothesis, 132
Evolutionary theory, 9–22
Darwin on, 10–12, 14–16
Hutton and Lyell, 12–14
pre-Darwinian thought, 9
pre-Darwinian thought, Age of Discovery, 10
pre-Darwinian thought, Age of Reason, 10–11
pre-Darwinian thought, Middle Ages, 9–10
theory of natural selection, 14–16
Evolution theory
evolutionary forces, 43–51
physical (biological) anthropology and, 1
primate evolution, 53–61

F

Fixity of Species, 9
Fleagle, J., 59, 90
Folate, 150
Forensic anthropology, 1
Founder effect, 45
Fox, E. A., 2
FOXP2 gene, 135
Frameshift mutation, 43
Franklin, R., 26
Frisancho, A. R., 148, 150

G

Galapagos Islands, 11–12
Galilei, G., 10
Gardner, B. T., 3, 78
Gardner, R. A., 3, 78
Gemmules, 23
Gene flow, 46, 131
Generalized skeleton, 67
Genes, 24
Genetic adaptation, 147
Genetic dating, 92
Genetic drift, 45–46
Genetics and heredity, 23–42. *see also* DNA
Mendel on, 4, 16, 23–26
molecular and population genetics, 26–32
population genetics, 32–33
regulatory genes, ix
Genotype, 24, 33
Genotypic ratio, 25
Geological time periods, defined, 89, 91. *see also* dating methods
Gibbons, 67, 74
Gingerich, P. D., 54, 55
Globalism, 151
Glutamic acid, 43
God hypothesis, 3
Goodall, J., 2
Gorillas, 67, 75–76, 79
Gould, S. J., 10, 13
"Grand Designer," 4
Great Chain of Being, 9–11
Grooming, 69
Guenons, 73, 79

H

Habitual bipedalism, 101
Haeckel, E., 121
Half-life time period, 92
Hall, Stephen, 135
Hamadryas baboons, 73
Hands, grasping, 68
Haplogroups, 138
Haploid, 27
Haplorhini, 70
Haplotypes, 138
Hardy, G., 32
Hardy-Weinberg equilibrium, 32–33
Heidelberg (Germany), 124
Heliocentric solar system, 10
Heterodontic teeth, 69
Heterozygote, 24
High altitude adaptation, 147, 150
Histoire Naturelle (Buffon), 12–13
HLA (Human Leucocyte Antigen System), 136
Holick, M. F., 149
Hominidae, 71
Hominims, 101, 111–114
Hominoidea, 71
Hominoids, 57
Homo erectus, 121–129
archaic *Homo sapiens,* 124–125
geography, 122–123
lithic technology, 125
morphology of, 121–122
Pithecanthropus erectus, 121
speech and language, 123
tools and subsistence, 123
Homogeneity, 45
Homo habilis, 110–114, 122
Homo neanderthalensis, 131–133, 134
Homo sapiens, archaic, 124–125
Homo sapiens neanderthalensis, 134
Homozygote, 25
Houle, A., 56
Hoyle, F., 3
Human adaptation, 147
Human biology, 2
Human osteology, 2
Human paleodemography, 2
Human paleogenetics, 2
Human paleontology, 1
Human paleopathology, 1
Hunting. *see* tool use
Hunting-gathering-scavenging, 123
Hutton, J., 12–13
Hybrids, 25
Hylobatidae, 71
Hyoid bone, 135
Hyperthermia, 149
Hypothermia, 148
Hypothesis, 2
Hypoxia, 150

I

Inheritance, 32
Inheritance of Acquired Characteristics, 14
Inheritance of variation, 15
Intelligent design creationism, 3
Isbell, L., 102
Isotopes, 91–92

J

Jablonski, N. G., 150
Japanese macaque, 73
Johnson, G. B., 15
Jones, F. W., 53–54
Jurmain, R., 9, 10

K

Kabwe (Zambia), 124
Karafet, T. M., 138
Kepler, J., 10
Knott, C. D., 2
Knuckle walkers, 75
Krause, J., 136
Krings, M., 135

L

Lactose intolerance, 29–30
Lamarck, J. B., 13–15
Larsen, C. S., 13, 15, 23–26, 57, 59, 89, 92, 109, 122, 134, 151
Law of independent assortment, 25
Law of segregation, 25
Law of superposition, 91
Law of Use and Disuse, 14
Lemurs, 69, 70
Levallois technique, 125
Lewin, R., 2, 78
Lightfoot, Rev. J., 12
Linden, E., 3, 78
Lindroth, S., 10
Lineage, 131
Linnaeus, C., 10, 14, 70, 101
Lion tamarin, 79
Lithic technology, 125. *see also* tool use
Locus (loci), 30
Loomis, W. F., 149
Lovejoy, A. O., 11
Lovejoy, C. O., 102
"Lucy," 105
Lyell, C., 12–13

M

Malaria, 44–45
Malaria-sickle cell hypothesis, 44
Male dominance, 77
Male provisioning hypothesis, 102
Malthus, T., 14
Mandrills, 73
Matsuzawa, T., 3
McCrossin, M., 57
McGrew, W. C., 2
McHenry, H., 102
McKie, R., 131
Medieval period, defined, 9–10
Megafauna, 139
Meiosis, 27–28
Melanin, 149
Melanocytes, 149
Mendel, G., 4, 14, 16, 23–26, 32
Mendelian ratio, 25
Mercader, J., 2
Mesozoic era, 89–90
Messenger RNA (mRNA), 28
Microevolution, 26–33
Miocene epoch, 56–58, 93, 101–102
Mitochondrial DNA (mtDNA), 132–133
Mitosis, 27–28
Modern humans, 131–146, 147–158
  adaptation by, 147–152
  agriculture, 137
  Asian ancestry for Native Americans, 138–139
  hypotheses for, 131–133
  Neanderthals, 133–136
  peopling of "New World," 137–138
  Proto-Cro-Magnons and Cro-Magnons, 136–137
Modern synthesis, 14, 26
Molecular anthropology, 2
Molecular clock, 92, 132
Molecular genetics, 26–32
Montagu, A., 9
Morgan, T. H., 26, 68
Mousterian complex, 135
Multiregional continuity hypothesis, 131
Mutation, 43

N

Nägeli, K., 25
Nariokotome boy *(Homo erectus)*, 122
Native Americans, Asian ancestry for, 138–139
Natural selection, 23
Natural selection theory, 14–16, 23, 43
Neanderthals, 133–136
Nelson, H. (check Newtons?), 9, 10
Newton, I., 2, 10
New World Monkeys, 55–56
New World monkeys, 70, 71
"New World," peopling of, 137–138
Ngandong (Indonesia), 124
Nocturnal activities, 68
Nonhoning complex, 103
Nonradiometric dating, 92
Nucleotide bases, 43
Numerical dating
  nonradiometric, 92
  radiometric, 91–92
Nystrom, P., 55, 56, 57, 67

O

Oldowan complex, 109
Old World monkeys, 56–57, 69, 70, 72
Oligocene epoch, 55
Oligopithecids, 55
Olorgesailie, 123
Omnivores, 151
Omomyids, 54–55
On the Origin of Species (Darwin), 11–12, 15
Orangutans, 67
Order, 70
Oreopithecus, 58
Organgutans, 74
Orrorin tugenesis, 103
Osteomalacia, 151
Ouranopithecus, 58
Out of Africa hypothesis, 131–133

P

Paleocene epoch, 54
Paleoindians, 137–138, 139
Paleozoic era, 89
Palmigrade quadripeds, 105
Pangaea, 89, 90
Parallel evolution, 56
Parapithecids, 55
Parental investment, 69
Patterson, F., 3, 78
Petralona (Greece), 125
Phanerozoic eon, 89
Phenotype, 24
Phylogenies, 90
Phylum (phyla), 70
Physical (biological) anthropology, 1–8
  defined, ix, 1
  meaning of being "human," 2–4
  subfields in, 1–2
Physiological adaptation, 147–148
Pigmentation, 149
Piloscene epoch, 93
Pithecanthropus erectus, 121
Plant domestication, 137
Platyrrhines, 55
Platyrrhini, 70
Pleiotrophic, defined, 32
Pleistocene, 57, 93
Pleistocene epoch, 93
Plesiadapiforms, 54
Pliocene, 57, 102
Point mutation, 43

Polygenic variation, 32
Polymorphism, 30
Pongidae, 71
Population genetics, 26–33
Postorbital bar, 54
Postorbital closure, 54
Potassium-40, 92
Pre-australopithecines, 102–105
Precambrian eon, 89
Prehensile tails, 71
Primate evolution, 53–61
  anthropoids, 55–59
  euprimates, 54–55
  hypotheses for primate origins, 53–54
  proprimates, 54
Primates, physical anthropology and, 1
Primates (living), 67–87
  biogeography of, 67
  defined, 67
  general characteristics (adaptations) of, 67–69
  taxonomic classification, 70–80
Primatology, 2
Proboscis monkeys, 73
Proconsulids, 56–57
Propliopithecids, 55
Proprimates, 54
Prosimians, 70
Proteins, 43
Protein synthesis, 28–30, 43
Proto-Cro-Magnons, 136–137
Pruetz, J. D., 2
Punnett square, 24

Q

Qafzeh fossils, 136
Quadramous adaptation, 74
Quadrupedal walkers, 75

R

Race, ix
Radiometric dating, 91–92
Raven, P. H., 15
Recessive traits, 24
Recombination, 28
Reduced sense of smell, 68
Regulatory genes, ix, 29
Reich, D., 136
Relative dating, 91
Relethford, J. H., 3, 11, 12, 32, 56, 89, 122, 134, 138, 151
Rhesus macaque, 73
Rhinarium, 68
Rickets, 150
Robust australopithecines, 108
Rodman, P., 102
Rumbaugh, D. M., 78

S

Sacculated stomachs, 72
Sahelanthropus tchandensis, 102
Sakolsky, J., 10
Savage-Rumbaugh, S., 2–3, 78
Schurr, T. G., 138
Scientific method, 2
Sedentism, 137
Sexual dichromatism, 74
Sexual dimorphism, 72
Shunkov, M., 135
Sickle cell anemia, 44–45
Sickle cell gene, 45
Silk, J. B., 23, 25, 26
Simons, E., 54
Sivapithecus, 58
Skhul fossils, 136
Smith, B. D., 90, 138
Smith, G. E., 53–54
Snowden, C. T., 3
Speciating, 131
Species, 131
Sponheimer, M., 111
Stabilizing selection, 44
Stable carbon isotopes, 113–114
Stable carbon-12 isotopes, 93
Stable carbon-13 isotopes, 93
Stable oxygen isotope studies, 93
Stasis, 10
Stereoscopic-binocular vision, 54, 67
Strata, 91
Strategiography, 91
Strepsirhini, 70
Stringer, C. B., 131
Struggle for existence, 15
Sussman, R., 54
Sutton, W., 26
Szalay, F. S., 55

T

Taxonomy, 70
Terrace, H. S., 78
Terrestrial quadrupeds, 72
Theories, defined, 2
Theory of the Earth (Hutton), 13
Thermoluminescence, 92
Tool use
  Cro-Magnons, 137
  Darwin on, 101
  evidence of, in Australopiths, 109–110
  Homo erectus, 123, 125
  Neanderthals, 135
Transcriptions, 28
Translation, 29
Trinkaus, E., 135
True-breeding, 23
Tyson, N., 2, 10

U

Ungar, P., 111
Uniformitarianism, 13
Use and Disuse, Law of, 14
Ussher, James, 12

V

Van Schaik, C. P., 2
Variation in fitness, 15
Vasoconstriction, 148
Vasodilation, 148
Victoriapithecids, 57
Victoriapithecus, 57
Visual predation hypothesis, 53, 54
Vitamin D, 149
Volcanic rock, 91
Von Seysenegg, E. T., 26

W

Walker, A., 57, 123
Wallace, A. R., 15
Watson, J., 26
Weinberg, W., 32
Wheeler, P., 102
White, T. D., 102, 105
Wolpoff, M. H., 131, 134

Y

Y-5 molar cusps pattern, 74
Young, T., 102

Z

Zhoukoudian, 123